Springer-Verlag Berlin Heidelberg GmbH

Antonino Cattaneo Silvia Biocca (Eds.)

Intracellular Antibodies:

Development and Applications

 Springer

Antonino Cattaneo, Ph. D.

Biophysics Sector
International School for Advanced Studies
(SISSA)
Via Beirut 2/4
34013 Trieste
Italy

Silvia Biocca, Ph.D.

Department of Experimental Medicine
and Biochemical Sciences
University of Rome „Tor Vergata"
Via di Tor Vergata 135
00133 Roma
Italy

ISBN 978-3-662-07994-2
Biotechnology Intelligence Unit

Library of Congress Cataloging-in-Publication Data

Cattaneo, Antonino, 1954-
Intracellular antibodies: developments and applications / Antonino Cattaneo, Silvia Biocca.
p.cm. – (Biotechnology intelligence unit)
Includes bibliographical references and index.
ISBN 978-3-662-07994-2   ISBN 978-3-662-07992-8 (eBook)
DOI 10.1007/978-3-662-07992-8

1. Immunoglobulins. 2. Immunization. 3. Cellular immunity. 4. Immunogenetics.
I. Biocca, Silvia. II. Series
[DNLM: 1. Immunity, Cellular–physiology. 2. Antibodies–physiology. 3. Antibody Formation–physiology.
4. Gene expression–physiology. QW 568.C3681 1997]
QR 186.7.C38 1997
616.07'98–DC21 97-20286 CIP
DNLM/DLC
for Library of Congress

© Springer-Verlag Berlin Heidelberg 1997
Originally published by Springer-Verlag Berlin Heidelberg New York in 1997
Softcover reprint of the hardcover 1st edition 1997

The use of general descriptive names, registered names, trademarks, etc. in this publication does not imply, even in the absence of a specific statement, that such names are exempt from the relevant protective laws and regulations and therefore free for general use.

Product liability: The publisher cannot guarantee the accuracy of any information about dosage and application thereof contained in this book. In every individual case the user must check such information by consulting the relevant literature.

Typesetting: Landes Bioscience Georgetown, TX, U.S.A.

SPIN:10631251   31/3111 - 5 4 3 2 1 0 - Printed on acid-free paper

Nec me animi fallit Graiorum obscura reperta
difficile inlustrare Latinis versibus esse,
multa novis verbis praesertim cum sit agendum
propter egestatem linguae et rerum novitatem;
sed tua me virtus tamen et sperata voluptas
suavis amicitiae quemvis efferre laborem
suadet et inducit noctes vigilare serenas
quaerentem dictis quibus et quo carmine demum
clara tuae possim praepandere lumina menti,
res quibus occultas penitus convisere possis.

Lucrezio Caro, De rerum natura (1,136-145)

PREFACE

The concept of exploiting the virtually unlimited repertoire of antibodies by ectopically expressing antibody genes for functional studies is ten years old. During this period, many developments have transformed this idea into a technology with broad potential. Now the promises, the problems and the perspectives can be more clearly identified. This volume addresses these issues by describing what has been achieved already, what could be conceivably achieved in the future and which problems require solutions and improvements. The volume is intended for people involved in different areas of research and therefore, where possible, we have tried to describe the background to the technology and to the applications.

The book reflects a personal view of how the field of ectopic antibody expression has evolved and of the directions in which it may evolve. Ideas are presented without constraints. The focus is on the general concepts and ideas, but methodological aspects are covered as well.

We are particularly grateful to our friend Roberto Sitia, whose help, experience and lively discussions were vital for many aspects of this work. Chapter 4 reflects in part his ideas, filtered through our own eyes and biases.

We feel privileged to have received, for many years, the support and guidance of Rita Levi-Montalcini, whose seminal experiments on immunosympathectomy have provided a germ which has developed into this work. We are very grateful to Rita for her support and friendship.

We would like to express our deep gratitude to Pietro Calissano, Director of the CNR Institute of Neurobiology (Roma), for his friendly and magnanimous support throughout many years, in particular in the initial phases of the development of this work.

A particular tribute goes to the researchers in our laboratories, who have contributed to this rapidly moving field in the past few years with their work, discussions, criticisms and high expectations, as well as to those who are starting now: Andrew Bradbury, Patrizia Piccioli, Francesca Ruberti, Anna Di Luzio, Nicola Gargano, Stefania Gonfloni, Luisa Fasulo, Michela Visentin, Massimo Righi, Lidija Persic, Marco Tafani, Samantha Messina, Alessio Cardinale.

Special thanks to Paola Pierandrei-Amaldi, for her friendly and enthusiastic scientific help, which established a turning point for us in the development of the intracellular antibody work.

The assistance of Samantha Messina and the cheerful collaboration of Vanessa Varnier in some phases of the preparation of this book are gratefully acknowledged.

AC is very grateful to Cesar Milstein and Michael Neuberger for thinking that the ectopic expression of antibodies was worth pursuing, and for allowing that pursuit to begin at the MRC Laboratory of Molecular Biology (Cambridge, UK). It was a pleasure to share lab 310 with Andrew and Michal. This book was partly written while AC was on sabbatical leave at the MRC Laboratory of Molecular Biology. AC wishes to thank Terry Rabbitts and Greg Winter, Joint Heads of the PNAC Division of the LMB, for the exciting opportunity.

SB thanks the Department of Experimental Medicine and Biochemical Sciences of the University of Rome "Tor Vergata" for assistance and support.

AC is grateful to Daniele Amati, Director of the International School for Advanced Studies (SISSA), for his enthusiastic encouragement and support.

The research mentioned in the text, from the authors' laboratories, has been supported by research contracts with NE.FAC within the National Research Plan Neurobiological Systems of the Ministero della Ricerca Scientifica e Tecnologica, Consiglio Nazionale delle Ricerche (CNR) (PF Biotecnologie, PF Ingegneria Genetica, PF Oncologia) INFM and S.I.R.S. srl. Part of the work has been intermittently funded by the Ministero della Sanita' (AIDS Project).

Finally, we would like to thank the staff of Landes Bioscience for their help and understanding.

A. Cattaneo and S. Biocca
Roma, 3 April 1997

CONTENTS

EDITORS

Antonino Cattaneo, Ph.D.
Biophysics Sector
International School for Advanced Studies (SISSA)
Trieste, Italy
Chapters 1, 2, 4-7, 10, 11

Silvia Biocca, Ph.D.
Department of Experimental Medicine and Biochemical Sciences
University of Rome "Tor Vergata"
Roma, Italy
Chapters 1, 2, 4, 5, 7, 10, 11

CONTRIBUTORS

Eugenio Benvenuto
ENEA, Dipartimento Innovazione
Divisione Biotecnologie e
 Agricoltura
Roma, Italy
Chapter 9

Andrew Bradbury, Ph.D.
Biophysics Sector
International School for Advanced
 Studies (SISSA)
Trieste, Italy
Chapter 3

Luisa Fasulo
Biophysics Sector
International School for Advanced
 Studies (SISSA)
Trieste, Italy
Chapter 10

Rosella Franconi
ENEA, Dipartimento Innovazione
Divisione Biotecnologie e
 Agricoltura
Roma, Italy
Chapter 9

Nicola Gargano
MRC Laboratory of Molecular
 Biology
Cambridge, U.K.
Chapter 10

Wayne A. Marasco, M.D., Ph.D.
Division of Human Retrovirology
 and Infectious Diseases
Dana-Farber Cancer Institute
Harvard Medical School
Boston, Massachusetts, U.S.A.
Chapter 8

Patrizia Piccioli
Institute of Neurobiology, CNR
Roma, Italy
Chapter 6

Francesca Ruberti
Biophysics Sector
International School for Advanced
 Studies (SISSA)
Trieste, Italy
Chapter 6

Paraskevi Tavladoraki
Universita' Roma Tre,
 Dipartimento di Biologia,
Roma, Italy
Chapter 9

CHAPTER 1

Intracellular and Intercellular Immunization

Antonino Cattaneo and Silvia Biocca

The Concept of Ectopic Antibody Expression for Intracellular and Intercellular Immunization with Antibody Genes

The virtually unlimited repertoire of antibodies is being exploited as a source of specific and selective chemical reagents/probes for both scientific and applicative purposes (including therapeutic), particularly after the advent of monoclonal antibodies. Antibodies or products derived from them by chemical or genetic engineering methods, can be delivered to a biological system as proteins or genes. This distinction is crucial to introduce the concept of intercellular and intracellular immunization with antibody gene.

Purified antibodies have been used for a long time in a very successful way to perturb gene function by injection in vivo. One prototypical example of this is provided by the classical experiment of "immunosympathectomy,"[1] whereby the injection of antibodies to the nerve growth factor in newborn rats was shown to induce cell death of sympathetic neurons. This experiment provided the formal proof that NGF was a growth factor required for the survival and differentiation of these cells.

Microinjection of antibody proteins in cells is a method that has been successfully used in basic research to inhibit intracellular gene function.[2,3] A variety of different phenotypic effects has given rise from microinjection of specifically directed antibodies (i.e., for example, the transiently reverted transformation of ras-activated fibroblast,[4] or the inhibition of neurite formation induced by NGF in PC12 cells[5] by microinjection), in both cases, of a purified anti-p21ras monoclonal antibody. However, the injected antibody may interfere with cytosolic or nuclear proteins but not with compartmentalized proteins in membranous districts such as endoplasmic reticulum (ER), Golgi or mitochondria. To interfere with proteins in the secretory pathway, microinjection of mRNA derived from hybridomas has been used and, the specific antibody was forced to interact with the antigen during its biogenesis.[6] Although suitable for the analysis of individual cells or groups of cells, the microinjection method of proteins or mRNA has several obvious limitations. Only short-term, fast biological responses can be studied because of antibody dilution, degradation and the need of repeated injections.

Intracellular Antibodies: Development and Applications, edited by
Antonino Cattaneo and Silvia Biocca. © Springer-Verlag and Landes Bioscience 1997.

The advent of monoclonal antibodies[7] and the availability of hybridoma cell lines secreting antibodies of predefined specificity allows isolation from these cells of not only the antibody molecules, but also the rearranged genes that encode for the antibody proteins. Vast repertoires of antibody genes are now even more readily available by exploiting the new phage display technology.[8,9] The availability of the genes coding for specific antibodies of interest led to the suggestion that their ectopic expression in a wide variety of nonlymphoid biological systems could be used to interfere with the corresponding protein.[10]

This book will focus on a description and critical discussion of this emerging technique based on the ectopic expression and subcellular targeting of recombinant forms of monoclonal antibodies, with the aim of blocking specific biological functions or of conferring new phenotypic traits (e.g., viral resistance), in biological systems where systematic genetic approaches are not feasible, for research purposes or in humans for therapeutic purposes. The technology is not limited to animal cells. In fact, one of its most important and potentially useful applications is in the field of plant biotechnology.

The development of the recombinant DNA technology has created a very successful and evolving field of applied research to produce new therapeutical antibodies and immunotoxins. This subject will not be discussed in this volume, except for those cases in which these are delivered as genes. It is clear that the entire field of therapeutical antibodies is limited to those antigens that can be accessed by extracellular delivery. The possibility of expressing antibodies "from within the cell" removes this limitation and greatly increases the range of antigens that may be the target of therapeutically useful antibodies. The delivery of antibody genes in vivo is a problem that requires many steps to be solved, an issue in common with other gene-based therapeutic approaches. The issue of antibody gene delivery will not be discussed in this book.

Ectopic Antibody Expression: Different Cells and Different Intracellular Compartments

The idea of ectopic antibody expression is to redirect the antibodies in space and time, to where and when the target protein recognized by the antibody is expressed. Thus, antibodies can be ectopically expressed (i) as secreted proteins (as they normally are in plasma cells) in cells that do not normally express them to interfere with extracellular antigens or (ii) as intracellular proteins targeted to different intracellular compartments to neutralize intracellular gene products.

Antibodies are normally secreted proteins which need several ancillary proteins for their correct assembly and folding during their biosynthesis in the endoplasmic reticulum of lymphoid cells (chapter 4). Following the demonstration that antibody assembly and secretion can occur efficiently also in nonlymphoid cells, it became clear that by engineering a suitable transcriptional control, the ectopic expression of secreted antibodies could be utilized to specifically perturb or interfere with the function of selected extracellular antigens in an otherwise intact tissue, as the nervous system (neuroantibodies), or organism as plants. Ectopic expression, in this case, refers to the expression of secreted antibodies in cells and in organisms that normally do not express immunoglobulins. This approach can be taken one step further by engineering the retargeting of immunoglobulins to different intracellular compartments. The intracellular immunization approach is based on the idea that antibody chains, if equipped with suitable targeting signals, could be targeted towards new ectopic intracellular sites. This can be obtained by taking advantage of the great wealth of information that has been accumulated on the different ways normal cellular proteins find their way inside the cell (chapter 5). This is based on the existence dominant and autonomous of specific localization signals that can be grafted onto other reporter proteins, conferring them a new intracellular localization (Fig. 1.1). The first decision that neosynthesized proteins make is whether they enter the secretory pathway or they remain in the cytosol. The presence or absence of an N-terminal leader sequence and the consequent synthesis on membrane bound or free ribosomes, is what determines this first choice. Proteins synthesized in the cytoplasm on free ribosomes will find their final location to other organelles or compartments by the presence of targeting signals. Proteins cotranslationally inserted across the endoplasmic reticulum (ER) membrane into the lumen of the ER undergo a hierarchically organized sequence of targeting decisions, some of which are also related to quality control mechanisms.

The rearranged genes coding for antigen specific antibodies can be equipped with suitable promoters and with suitable targeting signals to be expressed in different cells and different intracellular compartments.

Historical Overview

Early experiments in which total mRNA isolated from a hybridoma cell line producing an antibody against a viral protein was microinjected into heterologous cells,[6] have paved the way to the concept of a "monoclonal neutralization strategy" based on nucleic acid rather than protein. As a source of nucleic acid mRNA is not suitable for further engineering. The demonstration that antibody genes could be manipulated in vitro to produce new antibody forms and fusion proteins limited only by imagination,[11] showed the way to proceed. Table 1.1 shows an historical overview of the strategy of ectopic antibody expression to perturb gene products.

The prerequisites for the approach of ectopic antibody expression are: (i) the availability of sources for the isolation and selection of antibody encoding genes; (ii) the possibility of engineering modified antibodies and (iii) the use of gene transfer techniques for the particular biological system of interest. This approach combines and extends to stable expression what was previously possible only in a transient and limited way with microinjection of antibody protein or of antibody mRNA.

In order for this approach to be utilized, the following questions needed to be answered: (i) can nonlymphoid cells sustain the complex post-translational steps required to assemble a functional immunoglobulin? (chapters 4 and 6); (ii) can individual antibody chains be correctly targeted to the desired intracellular compartment? (iii) do the retargeted antibody chains fold correctly, so as to assemble with the cognate chain and preserve the antigen binding specificity? (iv) do the intracellular antibodies interact in vivo with the target antigen and neutralize its action? (chapter 7).

A positive answer to these questions has come over the past decade (Table 1.1) and constitutes the feasibility background on which the present success and potential of the approach is based. Many of the results quoted in the Table 1.1 will be discussed more extensively in the rest of the book.

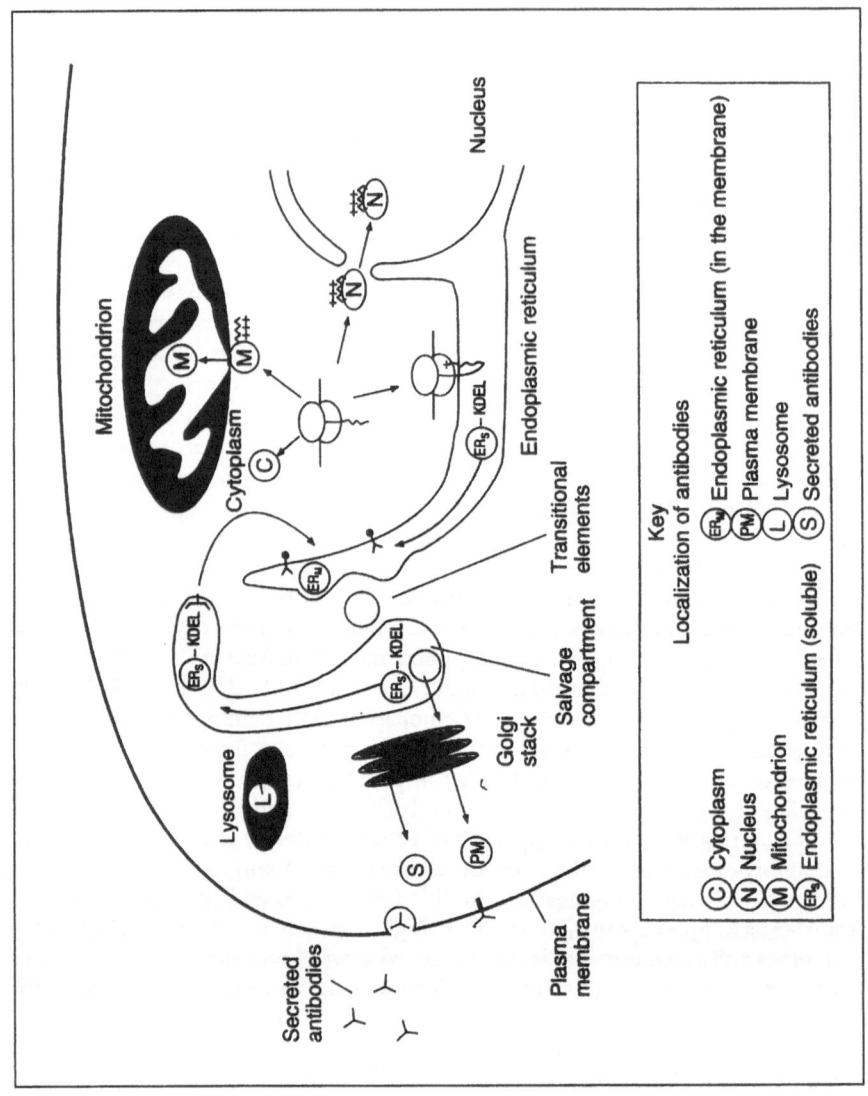

Fig. 1.1. Targeting of intracellular antibodies. The scheme illustrates how some of the known targeting signals (e.g., nuclear localization sequence, mitochondrial presequence and KDEL retention sequence) can be exploited to redirect the intracellular trafficking of antibodies in eukaryotic cells. Reprinted with permission from Biocca S and Cattaneo A, Trends Cell Biol 1995; 5: 248-252.

Table 1.1. *Ectopic antibody expression to perturb gene products: historical overview*

Block of transport of a membrane viral protein by mRNA hybridoma injection (6)
  Burke and Warren 1984

Ectopic expression of antibody genes to perturb extracellular molecules (10)
  Cattaneo and Neuberger 1987

Intracellular antibodies in yeast (13)
  Carlson 1988

Intracellular antibodies in mammalian cells (15)
  Biocca et al 1990

Antibody fragments in transgenic plants (16)
  Benvenuto et al 1991

Neutralization of p21ras activity by intracellular expression of ScFvs (17)
  Biocca et al 1993

Inhibition of viral replication in transgenic plants by intracellular expression of ScFvs (18)
  Tavladoraki et al 1993

Block of transport of HIV-1 gp120 protein by expression of ScFvs reduces HIV-1 infectivity (19)
  Marasco et al 1993

Inhibiting a neuropeptide by the expression of antibodies in the central nervous system of transgenic mice (20)
  Piccioli et al 1995

The term intracellular immunization that is used extensively in this book, together with the complementary one of intercellular immunization, has been introduced in a 1988 commentary by D. Baltimore,[21] to describe gene-based approaches used to confer cellular resistance to viral infection. The absence of antibody genes amongst the list of the candidate genes potentially exploitable for intracellular immunization is remarkable, notwithstanding the immunological vocabulary used to describe it. Today, the antibody based intracellular immunization strategy has acquired a prominent space in different research and applicative fields.

Applications in Research, Gene Therapy and Biotechnology

As outlined above, the potential of this approach is great for both research and applicative purposes since it is based on the virtually unlimited repertoire of the immune system. This is presently being further enriched by the availability of synthetic repertoires of cloned antibody genes (chapter 3). Unlike other gene-based methods to perturb gene functions in higher organisms, antibodies can selectively target nonprotein antigens or post-translational modifications of target proteins (including heterodimerization).

Applications in research are most valuable for those systems where a genetic approach is not feasible, and in those tissues where the complexity of the organization calls for detailed and subtle perturbation methods. The application to the nervous system is one which has been recently developed and will be specifically described in chapter 6.

The expression of intracellular antibodies is finding its more promising application in a gene therapy perspective towards the creation of cellular resistance to viral infection in human pathology (chapter 8).

Antibodies have also been ectopically expressed in plants. While this has been initially aimed at assessing the feasibility of using transgenic plants as an alternative expression system for the production of therapeutically important antibodies, subsequent studies have shown that, by the expression of intracellular antibodies, resistance to pathogenic viruses can be engineered in transgenic plants. This biotechnologically important field of application in transgenic plants is extensively described in chapter 9.

The efficiency of this strategy depends crucially on an efficient folding of antibody domains in ectopic environments. Antibody folding in *E. coli* is being actively studied (chapter 4). The requirements for an efficient folding of antibody domains will depend on the particular cell compartment in which the antibodies are expressed. The potential of designing selection schemes for the isolation of antibodies active in selected intracellular compartments will be discussed in chapter 10.

References

1. Levi MR, Angeletti PU. Immunosympathectomy Pharmacol Rev 1966; 18:619-628.
2. Graessmann A, Graessmann C, Mueller C. Microinjection of early SV40 DNA fragments and T antigen. Methods Enzymol 1980; 65:816-825.
3. Morgan, DO, Roth RA. Analysis of intracellular protein function by antibody injection. Immunol Today 1988; 9:84-86.
4. Feramisco JR, Clark R, Wong G et al. Transient revertion of ras oncogene-induced cell transformation by antibodies specific for amino acid 12 of ras protein. Nature 1985; 314:639-642.
5. Hagag N, Halegoua S, Viola M. Inhibition of growth factor- induced differentiation of PC12 cells by microinjection of antibody to ras p21. Nature 1986; 319:680-682.
6. Burke B, Warren G. Microinjection of mRNA coding for an anti-Golgi antibody inhibits intracellular transport of a viral membrane protein. Cell 1984; 36:847-856.
7. Koehler G, Milstein C. Continuous cultures of fused cells secreting antibody of predefined specificity. Nature 1975; 256:495-497.
8. McCafferty J, Griffith AD, Winter G et al. Phage antibodies: filamentous phage displaying antibody variable domains. Nature 1990; 348:552-554.
9. Winter G, Griffiths AD, Hawkins RE et al. Making antibodies by phage display technology. Ann Rev Immunol 1994; 12:433-455.
10. Cattaneo A, Neuberger MS. Polymeric immunoglobulin M is secreted by transfectants of non-lymphoid cells in the absence of immunoglobulin J chain. EMBO J 1987; 6:2753-2758.

11. Neuberger MS, Williams GT, Fox RO. Recombinant antibodies possessing novel effector functions. Nature 1984; 312:604-608.
12. Valle G, Jones J, Colman A. Anti-ovalbumin monoclonal antibodies interact with their antigen in internal membranes of *Xenopus* oocytes. Nature 1982; 300:71-74.
13. Carlson JR. A new means of inducibly inactivating a cellular protein Mol Cell Biol 1988; 8:2638-2646.
14. Hiatt A, Cafferkey R, Bowdish K. Production of antibodies in transgenic plants. Nature 1989; 342:76-78.
15. Biocca S, Neuberger MS, Cattaneo A. Expression and targeting of intracellular antibodies in mammalian cells. EMBO J 1990; 1:101-108.
16. Benvenuto E, Ordàs R, Tavazza R et al. 'Phytoantibodies': a general vector for the expression of immunoglobulin domains in transgenic plants. Plant Mol Biol 1991; 17:865-874.
17. Biocca S, Pierandrei-Amaldi P, Cattaneo A. Intracellular expression of anti-p21ras single-chain Fv fragments inhibits meiotic maturation of *Xenopus* oocytes. Biochem Biophis Res Comm 1993; 197:422-427.
18. Tavladoraki P, Benvenuto E, Trinca S et al. Transgenic plants expressing a functional single-chain Fv antibody are specifically protected from virus attack. Nature 1993; 366:469-472.
19. Marasco WA, Haseltine WA, Chen S. Design, intracellular expression, and activity of a human anti-human immunodeficiency virus type 1 gp120 single-chain antibody. Proc Natl Acad Sci USA 1993; 90:7889-7893.
20. Piccioli P, Di Luzio A, Amann R, et al. Neuroantibodies: ectopic expression of a recombinant anti-Substance P antibody in the central nervous system of transgenic mice. Neuron 1995; 373-384.
21. Baltimore D. Intracellular immunization Nature 1988; 335:395-396.
22. Biocca S, Cattaneo A. Intracellular immunization: antibody targeting to subcellular compartments. Trends Cell Biol 1995; 5:248-252.

CHAPTER 2

Current Methods for Genotypic and Phenotypic Knock-Outs in Mammalian Cells

Antonino Cattaneo and Silvia Biocca

As more and more genes are sequenced and cloned, the problem of identifying the function of the gene, what its protein actually does in a living cell, is of fundamental importance. The classical approach taken from genetics is that of inactivating the gene and studying what effect it has. The emerging field of "functional genomics" depends crucially on the use of genotypic or phenotypic knock-out methods, and on the development of high-through put screening procedures. The possibility of interfering in an efficient way with the function of a target gene has led to a number of potential therapeutic applications.

The function of a gene can, in principle, be inactivated, i.e., at the level of DNA, or of the corresponding mRNA or protein product. This chapter will review and compare current methods of inhibiting gene function in higher organisms for research or for therapy, with a particular focus on gene based methods.

Gene Targeting

While it has been possible to disrupt a gene by targeted insertion in organisms such as yeast or *Drosophila*, this has only been feasible in the mouse since the end of the 1980s. Two key technologies have facilitated this experimental system: the isolation of embryonic stem cells as permanent in vitro cell lines that can repopulate the blastocyst stage embryo[1] and the discovery that mammalian cells could recombine introduced vector DNA with a homologous chromosomal target, a process known as gene targeting.[2] Gene targeting by homologous recombination in embryonic stem cells allows the production of mice containing a deletion in a predefined gene of interest called gene knock-out technology.[3] Homologous recombination in mammalian cells is a rare event compared to nonhomologous recombination events. Methods have been developed for the selection of these rare events in embryonic stem cells.[4]

This is a powerful technique for the functional study of gene products, which is finding many applications in creating mouse models with defined genetic lesions. Conventional gene knock-out techniques allow engineering of gene deletions in

Intracellular Antibodies: Development and Applications, edited by
Antonino Cattaneo and Silvia Biocca. © Springer-Verlag and Landes Bioscience 1997.

all cell types. This regionally and temporally unrestricted genetic deletion may lead to severe developmental defects or premature death which can preclude analysis of postdevelopmental gene functions. If mutant mice complete development, interpretation of the experimental results often runs into two types of difficulties. First, global gene knock-out makes it difficult to attribute abnormal phenotypes to a particular type of cells or tissues. Second, it is often difficult to exclude the possibility that the abnormal phenotype observed in adult animals arises indirectly from a developmental defect. From the very beginning of development, the constitutive deletion of a gene may trigger compensatory changes that may mask the phenotype caused by the gene deletion.

Recent improvements of the gene knock-out technology allow spatiotemporal control of the genetic deletion. One method to accomplish cell-type or tissue-type restricted gene knock-out is to exploit the Cre/loxP system, a phage P1 derived site-specific recombination system in which the Cre recombinase catalyzes recombination between 34 bp loxP recognition sequences.[5] The loxP sequences can be inserted into the genome of embryonic stem cells by homologous recombination, such that they flank one or more exons of a gene of interest (called a "floxed gene"). It is crucial that the insertions do not interfere with the gene's normal expression. Mice homozygous for the floxed gene are generated from these embryonic stem cells and crossed to a second mouse that harbors a Cre transgene under the control of a tissue type- or cell type-specific transcriptional promoter. In progeny that are homozygous for the floxed gene and that carry the Cre transgene, the floxed gene will be deleted by the Cre/loxP recombination but only in those cell type(s) in which the Cre gene-associated promoter is active. It is clear that the extent of regulation of the gene deletion will be as good as the possibility of achieving a spatiotemporally controlled expression of the recombinase enzyme.

Mice containing T cell restricted gene knock-out were generated based on the above principles.[6] Very recently, this approach was extended to the production of mice in which region-specific gene knockout was accomplished in a highly restricted manner within postmitotic neurons of the forebrain, exclusively in CA1 pyramidal cells of the hippocampus.[7] In this work the aCaMKII promoter was used to drive the expression of the Cre transgene, but other transcriptional promoters with different brain-region specificity could be effective as well. This technology is very powerful as it combines the absolute inhibition obtained by gene knock-out with the possibility of spatiotemporal transcriptional regulation offered by engineering suitable promoter fusions; however, at the moment studies in another vertebrate species are not feasible with this technology and is limited to the mouse.

Some cells are more efficient than others in the homologous recombination events. In embryonic stem cells the homologous recombination process is efficien, and has led to the explosion of knock-out mice technology. For this reason, gene knock-out has had more success for studies on whole mice rather than in cell lines. From the cell biologist point of view, it is easier to generate gene deficient cell lines from a knock-out mouse than to knock out a gene directly in a cell line. The gene knock-out approach is very valuable for generating mouse models to study the function of the corresponding gene. It provides a very powerful tool for research purposes with the caveats mentioned above. For applicative purposes its potential appears to be more limited. From a therapeutic perspective, targeted insertion of a therapeutic gene in somatic cells, while a desirable goal, is still out of reach.

Antisense

The concept of antisense is based on the expression of nucleic acid complementary to a target mRNA to prevent the expression of the corresponding protein. In one approach, the complementary antisense nucleic acid molecules can be antisense RNA molecules expressed in the desired cell by gene transfer methods.[8] This method has provided both successes and failures in a rather unpredictable way, due to problems in secondary and tertiary RNA structure, stability and expression levels of the antisense RNA, design of optimal or suboptimal length, etc. Alternatively, antisense reagents can be designed as small oligodeoxynucleotides, able to be taken up from the extracellular medium by target cells.[9]

The development of antisense oligodeoxynucleotides, as therapeutic agents, relies on the ability of oligodeoxynucleotides to bind to a target disease-related RNA by Watson-Crick base pairing thereby inactivating it. The inhibitory effect of antisense on RNA translation can occur through a variety of mechanisms, one of which involves simple steric blockage of processes involved in RNA processing and translation (e.g., ribosomal movement along the mRNA). Recent studies suggest that a more important and efficient mechanism involves the irreversible cleavage of bound target RNA by endogenous RNAse H.[10] This has prompted an investigation of other RNA-degrading mechanisms including ribozymes, RNAse P and RNAse L.

The development of oligodeoxynucleotides (ODN) into effective antisense drugs requires that the following conditions be met: (i) metabolic stability; (ii) high affinity and specificity; (iii) ability to recruit RNAse H; (iv) cellular uptake at reasonable rates; (v) appropriate pharmacokinetic and pharmacodynamic behavior.

Phosphorothioate-modified ODN, in which one of the nonbridging oxygen atoms of the internucleoside phosphate group is replaced by sulfur, have progressed the furthest in clinical developments for a number of indications thanks to a substantial increase in resistance to nucleases compared to phosphodiester ODNs.[9] The quest for further improved ODNs has benefited from a combination of chemical modifications, structural considerations, and has led to second and third generation ODNs,[9] which still have met, however, with mixed successes and failures. Discrepancies between improved results in vitro do not correspond to improved cellular results. Noteworthy are peptide nucleic acids (PNA) and heterocyclic base modifications (C-5 propynyl pyrimidine).[11]

In general, genuine antisense effects are difficult to demonstrate formally and require many controls that are not always carried out in the published studies. Sequence specific non-antisense effects can be particularly misleading for in vivo studies.[12,13] Another example of a non-antisense effect that shows some sequence specificity is an alteration of Sp-1 transcription factor activity by certain modified oligonucleotides.[14] Conversely, there can be nonspecific (but antisense-mediated) effects.

The selection of ODN sequences is still largely empirical. Virtually any region of the RNA can be proven to be the most effective target of antisense, unfortunately, this is still unpredictable a priori: the whole mRNA needs to be scanned in order to locate the most effective (or even an effective) ODN.

Ribozymes

The discovery that RNA can act as an enzyme has led to the emergence of a new field aimed at exploiting the enzymatic activity of engineered RNA molecules (ribozymes) to inhibit gene expression by the catalytic cleavage of a target mRNA molecule. The prototype of this new class of molecules is the hammerhead ribozyme which was first discovered in plant viroids, small circular pathogenic RNAs. Interest in this natural class of RNA enzymes was greatly stimulated after it was realized that the intramolecular RNA cleavage reaction could be turned into an intermolecular reaction by dividing the RNA into separate ribozyme and substrate strands.[15]

The ribozyme cleaves its RNA substrate behind a nucleotide NUH triplet, where N is any nucleotide and H is U, C or A but not G. The most efficiently cleaved substrate contains a GUC triplet. The sequence specificity of cleavage is determined by the complementarity between the substrate and the binding arms of the rybozyme. Thus the ribozyme can function as a sequence-specific RNAse, which is able to cleave any RNA containing a triplet amenable to hydrolysis, targeted through complementary flanking sequences. Recently, the crystal structure of the hammerhead ribozyme was elucidated, leading to an improvement in our understanding of RNA catalysis and its relation to RNA three-dimensional structure.[16,17]

Although ribozymes have been administered exogenously, the main way of introducing them into cells is by gene transfer of an expression cassette. Many different parameters of vector design influence the expression levels and stability of the ribozyme, and have to be optimized in a ribozyme expression cassette, including the promoter (inducible, tissue-specific or constitutive), as well as the cis-appended sequences required for RNA capping, transcript termination and export from the nucleus to the cytoplasm. Sequences in the retroviral vector used for delivery have also been shown to affect the efficiency of ribozyme expression from its promoter.

Tissue specific or inducible promoters have been successfully used to drive ribozyme expression.[18] Both RNA polymerase II and RNA polymerase III promoters have been used successfully to drive constitutive ribozyme expression. Pol III transcribes a variety of small nuclear and cytoplasmic RNAs that are abundant in all cell types. Pol III promoter elements from genes encoding for tRNAs, the U6 small nuclear RNA and virally-encoded VA RNAs that are expressed to high levels in adenovirus infected cells have been used to drive the expression of ribozymes.

While the promoter controls the number of ribozyme transcripts produced in cells, many other factors that determine stability, localization and activity of ribozyme transcripts can be encoded in the transcription unit itself. Strategies for enhancing the accessibility of the target RNA, to pairing with the ribozyme by colocalizing the ribozyme close to the target mRNA, are being pursued following the initial study by Sullenger and Cech who demonstrated that colocalization of a ribozyme greatly improved the efficacy of the cleavage.[19] The ribozyme and the retroviral RNA encoding the target lacZ gene were colocalized via the retroviral dimerization domain, forcing copackaging of the retroviral transcript encoding the ribozyme with the target. Ribozymes can be targeted not only to cleave specific RNAs, but to also replace a defective portion of RNA with a functional sequence by targeted trans-splicing.[20] This represents a different application of ribozymes for which they may be uniquely suited. These molecules could potentially be adapted to correct a broad array of mutant transcripts. Despite the great potential of ribozymes, mainly due to their catalytic properties, the possibility of their targeting many aspects of the technology need to be optimized and refined.

Since base-pairing specificity determines the selectivity of a ribozyme for its target RNA, base-pairing mismatches or mutations adjacent to the site of cleavage will severely affect ribozyme efficiency. This problem is especially significant when designing ribozymes against genetically variable targets such as HIV. Moreover, ribozyme activity can be inhibited when ribozymes are inserted into complex transcriptional units that may fold into undesirable conformations. Finally, the intracellular ionic conditions may be very different from the ones determined for optimal enzymatic activity in vitro.

Dominant Negative Mutations

The function of a wild-type gene product can also be inactivated by the overexpression of an inhibitory variant of the same product, known as "dominant negative" mutation.[21] In general, dominant negative mutant proteins will retain an intact functional portion of the domain of the wild-type protein, but have the complement of this portion either missing or altered, so as to be nonfunctional. In this respect, some oncogenes are examples of naturally occurring dominant negative mutations. The design of a dominant negative mutant requires the domain structure or the functional structure of the protein to be known, therefore, the method is not sufficiently general. For some proteins of known structure (p21-ras) or of known domain organization (FGF-receptor) dominant negative mutants have provided an effective way of inhibiting their function.[22,23]

The use of dominant negative mutant derivatives require careful interpretation. Although the approach is designed to disrupt a wild-type gene function, the overproduction of an inactive product might have the opposite effect of increasing the activity of the wild-type protein. This might occur if the defective form of the protein titrates a cellular inhibitor.

Ectopic Antibody Expression

The use of ectopic antibody expression, described in this book, is in some way similar to the expression of dominant negative mutants of a gene, in that it acts as a dominant competitor at the level of the target protein, however. The method is more general: (i) it exploits the virtually unlimited repertoire of a class of structurally identical proteins, the antibodies; (ii) it allows targeting of functional subsets of a given protein, such as a pool of post-translationally modified versions of that protein; (iii) it allows targeting of pools of a protein with a particular subcellular localization; (iv) it can be used against nonprotein targets; (v) it allows, in principle, selective targeting of particular oligomeric combinations of a given protein (e.g., a particular heterodimeric complex, as opposed to a homomeric version of the same protein).

In more refined schemes, antibodies would not just be inhibitors; rather, effector functions could be engineered, coupled to the binding antibody moiety. The logic of the immune system, to which antibodies have a binding function provided by the variable regions, coupled to an effector function that is activated upon binding the antigen, could in principle be adapted to the particular needs of ectopic expression. Finally, ectopic antibody expression is not limited to a particular species, but has been shown to be effective in a wide spectrum of species of both animal and plant origin.

References

1. Evans MJ, Kaufman MH. Establishment in culture of pluripotential cells from mouse embryos. Nature 1981; 292:154-156.
2. Capecchi MR. Altering the genome by homologous recombination. Science 1989; 244:1288-1292.
3. Ramirez-Solis R, Davis AC, Bradley A. Gene targeting in embryonic stem cells. Methods in Enzymology 1993; 225:855-878.
4. Koentgen F, Stewart CL. Simple screening procedure to detect gene targeting events in embryonic stem cells. Methods in Enzymology 1993; 225:878-890.
5. Sauer B, Henderson N. Site-specific DNA recombination in mammalian cells by the Cre recombinase of bacteriophage P1. Proc Nat Acad Sci USA 1988; 85:5166-5170.
6. Gu M, Marth JD, Orban PC et al. Deletion of a DNA polymerase β gene segment in T cells using cell-type specific gene targeting. Science 1994; 265:103-106.
7. Tsien JZ, Chen DF, Gerber D et al. Subregion and cell type-restricted gene knock-out in mouse brain. Cell 1996; 87:1317-1326.
8. Izant JG, Weintraub H. Inhibition of thymidine kinase gene expression by anti-sense RNA: a molecular approach to genetic analysis. Cell 1984; 36:1007-1015.
9. Matteucci MD, Wagner RW. In pursuit of antisense. Nature 1996; 384:20-22.
10. Chiang MY, Chan H, Zounes MA et al. Antisense oligonucleotides inhibit inter-cellular adhesion molecule 1 expression by two distinct mechanisms. J Biol Chem 1991; 266:18162-18171.
11. Flanagan WM, Su LL, Wagner RW. Elucidation of gene function using C-5 propyne antisense oligonucleotides. Nature Biotech 1996; 14:1139-1145.
12. Krieg AM et al. CpG motifs in bacterial DNA trigger direct B cell activation. Nature 1995; 374:546-549.
13. Holt JT. A 'senseless' immune response to DNA. Nature Med. 1995; 1:407-408.
14. Perez JR. Sequence independent induction of Sp-1 transcription factor activity by phosphorothioate oligodeoxynucleotides. Proc Natl Acad Sci USA 1994; 91: 5957-5961.
15. Haseloff J, Gerlach WL. Simple RNA enzymes with new and highly specific endoribonuclease activities. Nature 1988; 334:585-591.
16. Scott WG, Klug A. Ribozymes: structure and mechanism in RNA catalysis. TIBS 1996; 21:220-224.
17. Sigurdsson ST, Eckstein F. Structure-function relationships of hammerhead ribozymes: from understanding to applications. TiBtech 1995; 13:286-289.
18. Rossi JJ. Controlled, targeted, intracellular expression of rybozymes: progress and problems. TiBtech 1995; 13:301-306.
19. Sullenger BA, Cech TR. Tethering ribozymes to a retroviral packaging signal for destruction of viral RNA. Science 1993; 262:1566-1569.
20. Sullenger BA, Cech TR. Ribozyme-mediated repair of defective mRNA by targeted trans-splicing. Nature 1994; 371:619-622.
21. Herskowitz I. Functional inactivation of genes by dominant negative mutations. Nature 1987; 329:219-222.
22. Feig LA, Cooper GM. Inhibition of NIH 3T3 cell proliferation by a mutant ras protein with preferential affinity for GDP. Mol Cell Biol 1988; 8:3235-3243.
23. Amaya E, Musci TJ, Kirschner MW. Expression of a dominant negative mutant of the FGF receptor disrupts mesoderm formation in *Xenopus* embryos. Cell 1991; 66:257-270.

CHAPTER 3

Recombinant Antibodies for Ectopic Expression

Andrew Bradbury

Antibody Structure

Antibodies or immunoglobulins are divided into 5 classes: IgA, IgD, IgE, IgG and IgM of which IgG is the most abundant in both mouse and human serum. The antibody molecule consists of four polypeptide chains, two identical heavy (H) chains and two identical light (L) chains held together by a combination of disulfide bonds and noncovalent interactions, as shown in Figure 3.1. Digestion of IgG with the proteolytic enzyme papain divides the antibody molecule into two identical antigen binding fragments, called Fabs, and a fragment called Fc. The polypeptide chains of immunoglobulins are composed of domains of similar structure, each consisting of two stacked layers of β sheets surrounding an internal space filled with hydrophobic amino acid side chains, with terminal exposed loops. They are termed either constant (C) or variable (V) on the basis of the degree of sequence variation amongst different antibody molecules, which is focused on the three, hypervariable, exposed loops at the top of the variable domains. The intervening strands of more rigid anti-parallel β sheet are termed framework regions and are highly conserved.[1] The hypervariable loops of a pair of VH and VL domains (H1, H2, H3 and L1, L2, L3) together form the antigen binding site. These loops, which vary in length and in sequence, are also known as complementarity determining regions (CDRs) due to their dominant role in determining the shape of the binding site and its specificity.

The VH domain is produced by somatic recombination of three gene segments: VH, D and JH. The VH gene segment encodes H1 and H2. H3 is formed by all three segments: the end of the VH, D and the beginning of JH. The VL domains are formed by a combination of two gene segments: VL and JL. VL codes for L1, L2 and the majority of L3.

Analysis of the relationship between the sequence and three-dimensional structure of the antibody combining sites[2-4] revealed that, except for H3, the other loops have one of a small number of main-chain conformations or canonical structures. The canonical structure formed in a particular loop is determined by its size and the presence of certain residues at key sites in the loop and in framework regions.

Intracellular Antibodies: Development and Applications, edited by
Antonino Cattaneo and Silvia Biocca. © Springer-Verlag and Landes Bioscience 1997.

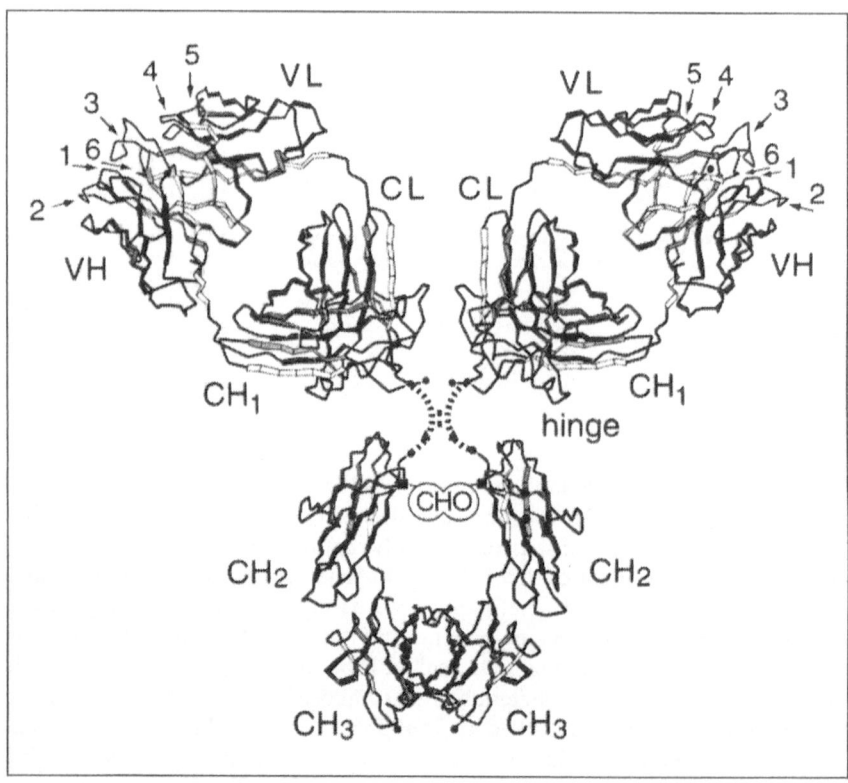

Fig. 3.1. Antibody structure. In the IgG molecule (depicted), the H chain consists of a variable domain (VH) and constant domains CH1, CH2 and CH3, whereas the L chain is comprised of sole variable (VL) and constant (CL) domains. The loops in the variable domains which contact to the antigen are labeled 1 to 6. The H chain contributes loops 1, 2 and 3 and the L chain 4, 5 and 6.

A wealth of structural information is now available that has provided insight into the precise nature of antigen-antibody interactions (for a review see refs. 5 and 6). Antigen binding sites in the antibody molecules can be divided into 3 general classes: clefts (which bind small molecules such as haptens and steroids), grooves (peptides and DNA) and planes (proteins). The surface area buried in the interaction can vary depending on the type of binding site and is 160-350 $Å^2$ for clefts, 400-600 $Å^2$ for grooves and 700-900 $Å^2$ for planar binding sites. Not all the CDR loops are necessarily involved in contacts with the antigen and in certain antibody-antigen complexes some framework residues interact directly with the antigen.

The enormous diversity of antigens requires a variety of different antibody specificities. The diversity of antigen binding sites on antibodies is produced in the immune system by two processes: combinatorial assembly of the primary repertoire (also including imprecisions in the joining of the different segments during recombination)[7] and antigen-dependent somatic hypermutation (for a review see refs. 8 and 9).

The two diversification processes, combinatorial rearrangements and somatic hypermutation, appear to be complementary, recent analysis having shown that assembly of the repertoire focuses diversity to the center of the binding site, whereas somatic mutation spreads diversity to regions at the periphery of the binding site that are highly conserved in the primary repertoire.[10]

Affinity and Avidity

The antigen-antibody interaction is quantitatively described by parameters such as the affinity constant and kinetic binding constants. These can be measured by a variety of different biophysical methods (for a review see ref. 11). If one considers a bimolecular reaction in aqueous solution involving a monovalent antibody and antigen

$$Ab + Ag = Ab\,Ag$$

at equilibrium, the concentrations of the three species are dictated by the equation

$$K_d = [Ab][Ag]/[Ab\,Ag]$$

where K_d is the dissociation constant, [Ab], [Ag] and [Ab Ag] are the concentration of the free antibody of the free antigen and of the antigen-antibody complex, respectively.

The affinity constant K_a is simply defined as

$$K_a = 1/K_d = k_{on}/k_{off}$$

where k_{on} and k_{off} are the kinetic association and dissociation constants respectively.

The affinity constant K_a describes the "strength" of the interaction between antigen and antibody in solution. Whenever one of the binding partners is present in multiple copies on a support that prohibits free diffusion (e.g., antigens on a cellular membrane or on a microtiter plate), the observed affinity constant corresponds to the true one only if the binder in solution is monovalent. Under the same conditions, a multivalent binder (e.g., an IgG) displays a higher apparent affinity (the "functional affinity" or avidity) by virtue of rebinding effects and chelate binding. This apparent affinity constant is not a universal thermodynamic constant, but is dependent on the antigen density on the solid support and provided the geometric considerations can be evaluated, they can be estimated.[12] While avidity effects (rebinding effects and multivalent binding) are to be avoided to obtain "true" affinity constants, they form the basis of important biological and technological processes, thereby, exploited to stabilize or strengthen binding interactions.

It should also be noted that the K_d value for a given antigen-antibody interaction, as determined in aqueous solution in vitro, may be very different from that effectively present in an in vivo situation due to crowding effects resulting in competitive interactions by other proteins and cell components. A comparison of affinity measurements for a given antigen-antibody pair, in vitro and in vivo, would be very informative.

Cloning of V Regions for Intracellular Expression

Intracellular immunizaton as a method for the inhibition of the function of cellular molecules requires genes encoding antibody binding regions as its raw material. Until recently, these would have invariably come from hybridomasbut, a new technique involving the display of antibody binding regions on the surface of filamentous phage (phage display) has recently been developed. Once generally available, is likely to become the method of choice for the isolation of antibodies in the future. This technique is particularly useful in intracellular immunizaton as the isolation of the antibody binding specificity is simultaneous with the cloning of the V regions encoding that specificity, thus eliminating one of the more troublesome steps in the use of intracellular immunizaton. This does not diminish the importance of monoclonal antibodies obtained in the classical way. So many useful specificities are presently available that for some time to come, antibody V regions used in intracellular immunizaton will be mainly derived from hybridomas. In this chapter, the principles behind the methods used to isolate antibody V region genes for use in intracellular immunizaton, first from hybridomas and secondly from phage display libraries will be described.

Cloning Hybridoma V Regions

The practical details involved in the cloning of hybridoma V regions has recently been reviewed in some detail.[13] In theory, the cloning of antibody V regions from hybridoma mRNA would seem to be a relatively straightforward procedure: immunoglobulin mRNA accounts for a large proportion of the mRNA in a hybridoma cell and the V regions encoding the binding specificities are relatively conserved.[14] It has been found, notwithstanding the theoretical ease of cloning, that the isolation of V region genes from hybridomas by commonly used PCR techniques can sometimes be extremely difficult,[15] the reasons for which will be discussed later. Four broad methods can be used to isolate V regions from hybridomas:

1. V region (or Fab) PCR
2. RACE (Rapid Amplification of cDNA Ends) or oligoligation PCR
3. cDNA cloning
4. phage display

The quickest method, when it works, is to amplify the hybridoma V regions using primers in or around the V regions, assemble them into single chain Fv (ScFv) fragments,[16,17] express them as soluble secreted proteins in bacteria and test them for binding activity. Alternative strategies which can also use V region PCR as the initial cloning procedure include the expression on the surface of phage (see below), followed by selection for binding activity or direct expression in mammalian cells as complete antibodies.

When V region PCR fails, RACE or a similar technique, oligoligation PCR, using conserved primers in the immunoglobulin constant regions is usually successful. If this method also fails, classical cDNA cloning using immunoglobulin constant region probes, is a method which is slower and labor intensive but invariably successful.

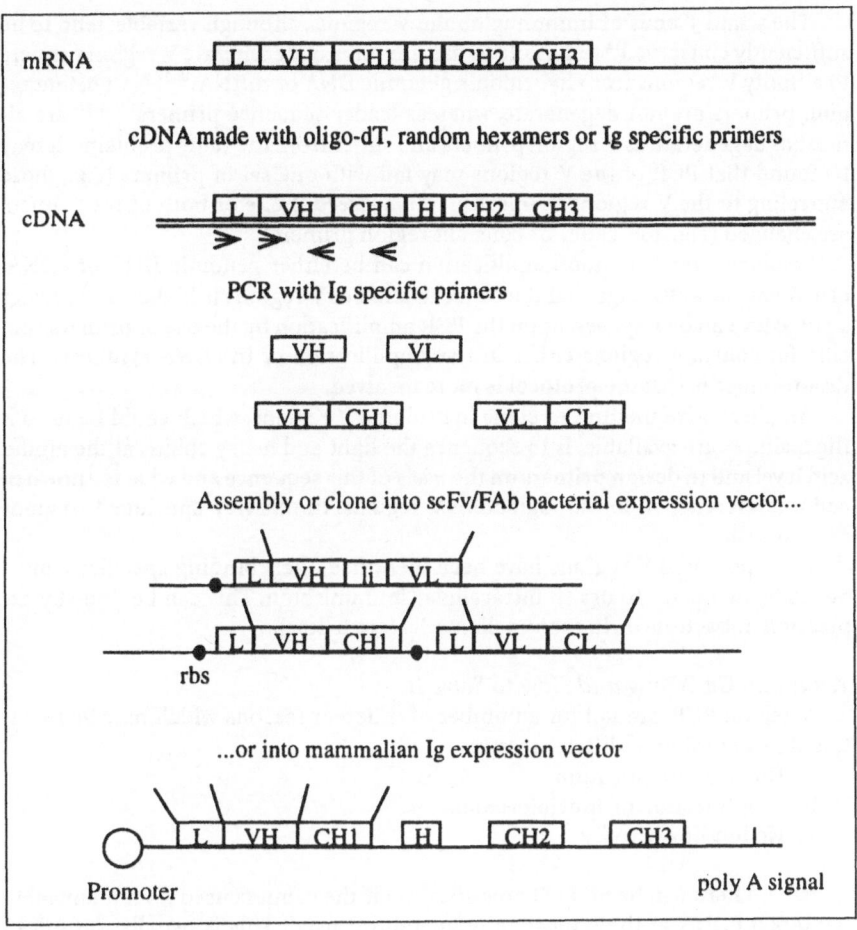

Fig. 3.2. Cloning of V regions. After cDNA has been made using any one of a number of different primers (oligo-dT, random hexamers or Ig specific primers), V region cloning by PCR may be carried out with primers recognizing the 5' end of the leader or V region and the 3' end of the V region or constant region, as indicated in the figure. The amplified V or V/C regions may then be either PCR assembled or cloned into bacterial or mammalian expression vectors.

1. V Region PCR

To isolate V regions from a hybridoma using V region PCR or its derivatives, one must use a 5' primer which anneals at the 5' end of the V region or upstream of it (e.g., in the leader sequence) and a 3' primer which anneals at the 3' end of the V region or downstream of it (e.g., anywhere in the constant region). Any 5' primer may be used with any 3' primer with the result that a large number of different V region derivatives can, in theory, be isolated (see Fig. 3.2). In general, two antibody forms are most commonly isolated: V regions[18,19] or V+CH1/V+CL regions.[20]

The 5' and 3' ends of immunoglobulin V regions, although variable, tend to be sufficiently conserved[14] to be able to use mixtures of degenerate V region primers to amplify V regions from hybridoma genomic DNA or mRNA.[18,21-23] Constant region primers are not degenerate, whereas leader sequence primers[19,24-26] are almost as degenerate as V region primers and suffer from the same problems. It may be found that PCR of the V regions may fail with one set of primers (e.g., those annealing to the V regions themselves, but succeed if one or both of the primers are changed (e.g., for leader or constant region primers).

Templates for V region amplification can be either genomic DNA or cDNA. cDNA has the advantage that the concentration of V regions is higher, and greater stringency can be imposed upon the PCR amplification by the use of primers specific for constant regions either in the amplification or in cDNA synthesis. The disadvantage is that the protocol is more involved.

An alternative method to clone hybridoma V regions, which could be used if the facilities are available, is to sequence the light and heavy chains at the amino acid level and to design primers on the basis of this sequence and what is known of codon preference in immunoglobulin V regions. In this way, spurious V regions are avoided.

Once potential V regions have been identified, their binding specificity must be confirmed prior to use in intracellular immunizaton. This can be done by expression in bacteria or in mammalian cells (see below).

What Can Go Wrong and How to Solve It

V region PCR can fail for a number of different reasons which may be manifested in a number of different ways:
   a. No PCR amplification
   b. Amplification of spurious bands
   c. No binding activity

a.   There will be no PCR amplification if the primers used do not anneal to the target cDNA at the annealing temperatures used. This is usually a result of either somatic mutation within the region recognized by the primer, or the presence of a V region or leader sequence which is not recognized by the primers used. This is more likely to be a problem when somatic mutation is accentuated as occurs in hybridomas derived from mice immunized over a long period. The solution to this problem is to either use RACE (see below) or alternatively to change one or both primers. Constant region primers annealing at either end of the first constant region, are less likely to suffer from this problem especially if the isotype of the antibody is known and specific primers are chosen accordingly.

b. The amplification of spurious bands is a problem encountered frequently and often not discovered until one is testing the binding specificity of the cloned antibody. One may be alerted to the presence of spurious V regions following the fingerprinting of the amplified V regions (using restriction enzymes which have four base recognition sequences), when the sum of the fragments obtained will exceed 300 base pairs. All amplified V regions should be at least, cloned identified and preferably sequenced. The extra V regions not encoding the binding specificity may be derived from: (i) V region derived from the parental myeloma; (ii) more than one VH or VL region present in the hybridoma; (iii) nonproductive V gene rearrangement and (iv) contaminant V regions.

Ideally, hybridomas would contain only the functional heavy and light chains which encode the binding specificity of the hybridoma. It is sometimes found that more than one functional chain is present. In many cases this is due to the presence of a functional light chain found in the myeloma fusion partner. These light chains are good substrates for V region PCR and will often be amplified in preference to the hybridoma light chain providing the specificity. This problem is dealt with more easily if one is aware of the sequence of the partner chains carried by the various fusion myeloma partners.[14] Where the myeloma partner is known to contain a nonproductive rearrangement of known sequence, the chains can be eliminated by using a specific oligonucleotide and RNAse H treatment.[27] In other cases, the presence of more than one chain indicates a mixed hybridoma which contains more than one clone or the result of the fusion of more than one spleen cell to the myeloma partner during the fusion process. It is clear that in the former case, further subcloning of the hybridoma should correct the problem. The additional V regions are intrinsic to the hybridoma and have to be dealt with by alternative means in the latter. As hybridomas can lose extra chromosomes, further subcloning can sometimes be helpful in this case too. Whatever the cause, the presence of extra functional V regions is a problem which can only be dealt with by the testing of each individual VH/VL combination for binding activity (unless parental myeloma V regions can be identified and thus eliminated). This can be done in a systematic fashion in bacteria or in mammalian cells, or alternatively by phage display (see below) when it is hoped that all possible VH/VL combinations are randomly produced and can be selected for on the basis of binding activity.

In addition to extra functional V regions, the PCR reaction may amplify nonfunctional V genes. These being the results of VDJ joining and N region diversity which do not reconstitute an in frame V region. As such nonfunctional rearrangements do not undergo somatic mutation, they are often better PCR substrates than the V region desired and are preferentially amplified.

In addition to the 'biological' reasons described above, extra V regions may result from contamination of the PCR by V regions which have been previously cloned in the laboratory. They are best avoided by careful negative controls for all PCRs and the use of dedicated pipettes and laboratory space for performing PCR.

c. After V regions have been cloned and expressed, either as ScFv molecules or full length immunoglobulins, it may be found that there is no binding activity for the antigen. In addition to the reasons given above, there are two other reasons which may cause this problem. It has been found that in some cases, the ScFv form of a monoclonal antibody does not bind the antigen. This may be due to a slight change of antibody binding site structure, to interference of the antibody binding site by the ScFv linker, or due to the formation of ScFv aggregates, and can be solved by expressing the full length immunoglobulin in mammalian cells. Another reason why the correct V regions may not show binding activity when expressed, either in the ScFv or full immunoglobulin forms, is because important amino acids involved in the binding activity have been changed by the PCR primers (Hoogenboom personal communication), a problem which can be overcome by using primers external to the V regions.

2. RACE and Oligoligation PCR

Given the variability of V regions and leader sequences, it is clear that a method to clone V regions independent of their sequence would be useful. Such a method is a modification of the RACE procedure[28] which was developed to clone the V genes of a monoclonal antibody recalcitrant to cloning by V region PCR.[15] This method involves a first strand cDNA synthesis using a primer specific for the Ig isotype of the hybridoma annealing within the hinge region for the heavy chain, addition of a polyA tail using terminal transferase, and amplification using one primer which anneals to the CH1 domain upstream of the hinge region and another which anneals to the 5' polyA tail (illustrated in Fig. 3.3). A similar method is used for the light chain. A similar technique, working on a similar principle, is oligoligation PCR,[29] in which instead of the addition of a polyA tail at the 5' end of the cDNA, a specific oligonucelotide is ligated using RNA ligase. PCR amplification of the immunoglobulin then occurs using the CH1 primer and another which is reverse and complementary to the oligonucleotide ligated to the end of the cDNA (see Fig. 3.3).

The amplification product obtained in both of these methods is one which contains: 5' untranslated leader, secretory leader sequence, V region and a variable amount of CH1 depending upon where the 3' primer was located. Although such a product cannot be expressed in bacteria (no bacterial ribosome binding site is present) unless the V (or V+C) region is reamplified and cloned into an appropriate bacterial expression vector, it can be expressed directly in most generic mammalian expression vectors, as well as those specifically designed for immunoglobulin expression.[30] If RACE has been used because no amplification product was obtained with V region PCR, the subsequent reamplification of the V region using V region primers can be done, but annealing temperatures as low as 37° may be required to overcome mutations found in the primer binding sites.[15]

A variation of this method, also involving the use of C region primers, followed by biotin-cDNA capture on streptavidin-coated magnetic beads and one-side PCR, has been described.[31]

3. cDNA Library Cloning

Before the days of PCR, immunoglobulins were cloned from hybridomas using classic cDNA synthesis techniques. Such methods, although somewhat time consuming, are invariably successful if one starts with mRNA of high quality derived from the hybridoma of interest and use constant region probes. The use of some specialist lambda vectors permits the rescue of plasmids which direct the expression of the cloned protein in mammalian cells without further subcloning.

4. Phage Display

Hybridoma V regions can be cloned using phage display. This is analogous to the method described below. The only difference is that instead of selecting for antigen binding activity from a large phage antibody library, one is selecting from a restricted library made from the hybridoma mRNA. This can be in either the Fab[20] or ScFv[32] format, and there are advantages and disadvantages for each. The ScFv format is generally easier to use, but the Fab format avoids the problems occasionally found with nonbinding ScFvs described above.

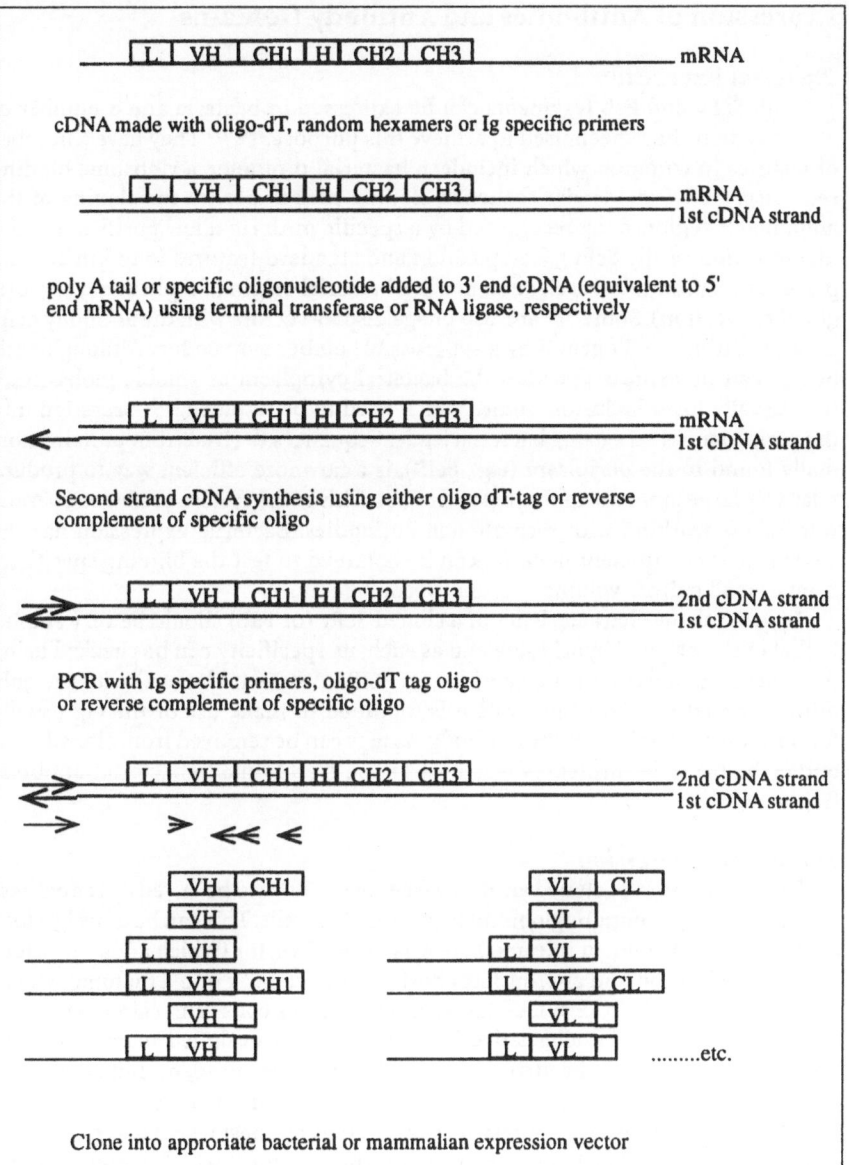

Clone into approriate bacterial or mammalian expression vector

Fig. 3.3. Using RACE to clone V regions. When V region primers are unable to amplify the correct V regions, RACE may be performed using general primers independent of the V region sequence being amplified. After the first cDNA strand has been synthesized, it is tagged using either terminal transferase to add a tail of As or RNA ligase to add a specific oligonucleotide (both indicated by the bold arrow in the figure). PCR is then performed using any combination of the primers indicated in the figure, the product of which can be cloned directly into mammalian or bacterial expression vectors.

Expression of Antibodies and Antibody Domains

Bacterial Expression

Both ScFv and Fab fragments can be expressed in bacteria and a number of vector systems have been used to achieve this purpose.[20,33,34] They have a number of features in common which include: a bacterial promoter, a ribosome binding sequence, a bacterial leader sequence, cloning sites to permit the cloning of the amplified V region, a tag recognized by a specific mAb (to allow purification and identification of the ScFv), a stop codon, and standard features found in cloning plasmids (e.g., antibiotic selection marker, bacterial origin of replication, M13 origin of replication). Some[20,34] are also phage display vectors with the antibody gene separated from the Ff gene 3 by a suppressible amber stop codon. Although antibodies can be expressed within the bacterial cytoplasm as soluble molecules,[35] they usually form inclusion bodies which need to be refolded.[36,37] Secretion into the periplasmic space using bacterial leader sequences derived from proteins normally found in the periplasm (e.g., pelB), is a far more efficient way to produce relatively large amounts of Fvs, ScFvs or Fabs. This is the system preferred by most researchers working with recombinant antibodies. Bacterial expression has the advantage that sufficient material can be obtained to test the binding specificity from a small culture volume after an overnight culture.

In general, the characteristics of a cloned ScFv (or Fab) should be very similar to that of the original hybridoma and as such, its specificity can be checked using the same system used for the hybridoma (ELISA, Western blot etc.) with the only difference that the detection system is modified to make use of the tag usually found at the C terminus of the antibody. As tags can be removed from cloned antibodies by bacterial proteases it is best to use freshly made bacterial antibody fragments.

Eukaryotic Expression

The specificities of cloned antibody fragments can also be tested by the expression of full length immunoglobulins in mammalian cells. This can be done by cloning the V regions into an intermediate M13 derived vector (which adds the leader sequence, its intron and appropriate sites) prior to cloning into a mammalian expression vector which provides the promoter, the Ig constant regions and poly adenylation site.[18,38] An alternative is to use specially designed mammalian expression vectors which, by virtue of the restriction sites used, permit single step cloning of the V region directly into the expression vector.[30] Once the V regions are cloned into one of these Ig expression vectors, they are best tested using transient transfectants in COS cells, as establishing stable cell lines which secrete enough antibody is too long a procedure to be used routinely for screening. Two to three days after transfection, sufficient antibody is produced (up to 4 μg/ml) for most analytical purposes (for a review see ref. 39).

When using expression in mammalian cells, one can choose to reconstitute the original mAb isotype or change it for another which has different properties. The expression vectors mentioned above use human constant regions, thus directing the synthesis of a hybrid mouse/man immunoglobulin with mouse variable regions and human constant regions.

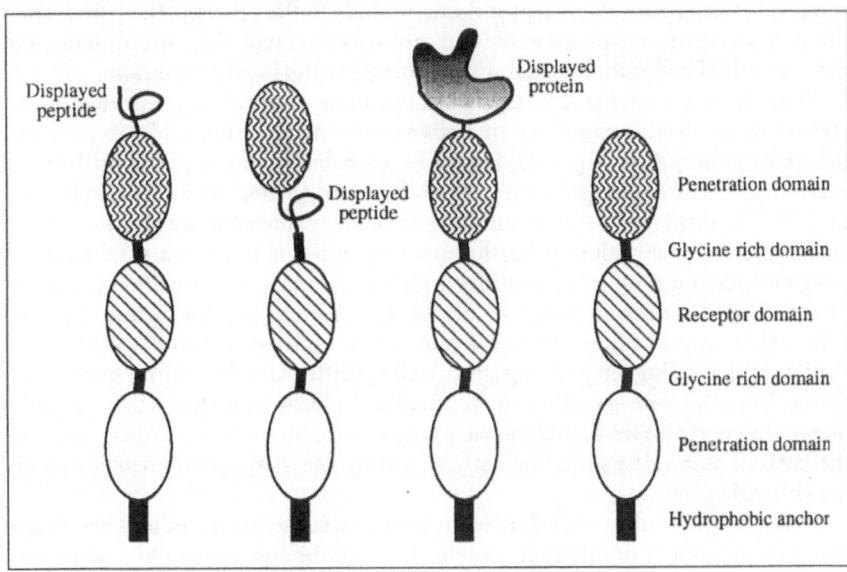

Fig. 3.4. The structure of p3 and sites used for display. p3 is composed of a number of different domains. The receptor domain is involved in binding to the F' pilus, the penetration domain is involved in the entry of the phage into the bacteria. The hydrophobic anchor anchors the protein within the bacterial membrane while the glycine rich domains which separate the different domains serve an unknown function. Peptides have been displayed between the penetration and receptor domains as well as at the N terminus of the protein. Proteins have only been displayed at the N terminus of p3 and this is now the site of preference for display.

Phage Display Antibody Libraries

As described above, antibody binding regions are likely to be isolated using phage display libraries in the future. This is a procedure which allows the selection of a protein and the gene encoding it on the basis of its binding specificity and is particularly suited for the isolation of antibodies.

Filamentous phage are long thread-like single stranded DNA phage which infect bacteria via sex pili. The best known, extensively used in molecular biology for producing single stranded DNA, are the Ff phage, a group of three independently isolated phages (M13, fd and f1) so similar at the DNA level that they can be considered to be minor variations of the same phage. These infect via the F pilus, inducing pilus retraction after binding to the tip and passing their DNA into the host bacteria (a process requiring both phage and host proteins) where it is replicated by a rolling circle mechanism. Unlike other phage which propagate by lysis of their hosts, filamentous phage do so by secretion, an unusual mechanism which slows bacterial growth (resulting in turbid plaques in top agar) and releases approximately 100 phage particles per division.

The phage protein which binds to the pilus tip is p3. This is present in three to five copies and consists of a number of different functional domains[40-42] (Fig. 3.4): the leader sequence directs p3 to the periplasmic space and is proteolytically removed after export, the receptor domain binds to the pilus tip, the penetration domain is essential for phage entry and probably interacts with host proteins, and the C terminal hydrophobic domain anchors p3 to the inner membrane.

It has been found that if foreign DNA encoding a peptide A is inserted downstream of the leader sequence or between the penetration and receptor domains of gene 3 (see Fig. 3.4), it will be translated and exposed within the mature p3 without compromising the ability of p3 to mediate infection via the F pilus.[43] This displayed peptide will be sufficiently exposed to be able to bind an anti-A monoclonal antibody. Furthermore, if anti-A is fixed to a solid support, phage displaying A or other peptides which mimic it, can be purified from a library of billions of such phage all displaying different peptides within their p3 (Fig. 3.5). This procedure involves a number of cycles of binding, washing and elution, with binding phage being specifically purified after about four or five such cycles. Since the gene encoding any particular chimeric p3 is always present in the phage which expresses that chimeric p3, any procedure which purifies phage on the basis of chimeric p3 binding activity will by necessity, simultaneously purify the chimeric gene 3.

While initial studies were done with peptides, subsequent studies have shown that a large number of different proteins (e.g., antibodies, growth hormone, proteases and their inhibitors, extracellular domains of membrane proteins such as CD4, zinc finger transcription factors and enzymes such as ampicillinase) can be displayed at the N terminal display site (Fig. 3.4) of p3 (see ref. 44).

Phage Antibodies

Antibodies, in either the ScFv or Fab format, have been one of the most successfully displayed proteins[34,45-48] with libraries of up to 6×10^{10} different members now having been displayed on phage using an in vivo recombination procedure based on the lox P system.[49] This allows the isolation of antibodies against virtually any antigen in a relatively simple in vitro procedure once a library has been obtained. As already described, concomitant with the isolation of an antibody will be the cloning of its gene, a result which avoids all the problems of V region cloning described above. The use of phage display to isolate antibodies is particularly useful for antigens which are either nonimmunogenic, or which cannot be produced by hybridoma cells for biological reasons (e.g., anti-BiP antibodies which are normally retained within the endoplasmic reticulum[32]). Phage display can also be used to isolate antibodies that have specificities which differ by a single amino acid.[50] Libraries have been created from immunized mice for the isolation of antibodies against specific antigens[20,51] (as an alternative to making hybridomas by traditional methods), as well as from unimmunized mice,[52] from unimmunized chickens[53] (which use gene conversion to generate diversity), from naturally immunized humans (to isolate antibodies against human pathogens[54,55]) or from deliberately immunized humans[56] (to isolate anti-melanoma antibodies). Single pot libraries, from which it has been possible to isolate antibodies against many different antigens, have been created from unimmunized humans[57] or synthetically from cloned human V genes.[32,49,58] Although irrelevant for intracellular immunizaton per se, phage display is also the simplest way to obtain human antibodies. This may be important if one is using intracellular immunizaton as a therapeutic measure where one would like to reduce the immune response to the intracellularly expressed immunoglobulin to a minimum.

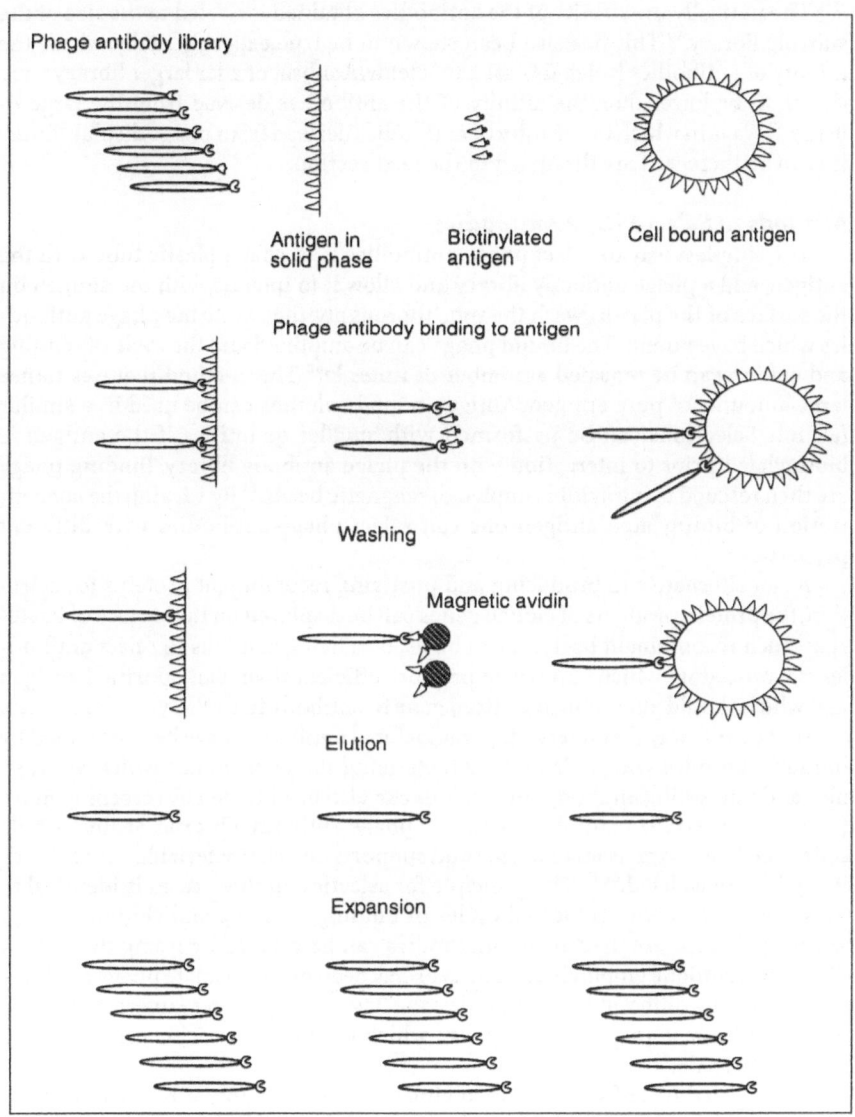

Fig. 3.5. Selection strategies. Specific phage antibodies recognizing an antigen of interest can be selected in solid phase where the antigen is coupled to a plastic surface; in solution, where the antigen is biotinylated; or on the surface of cells (bacterial or eukaryotic). All such selections involve recursive cycles of washing, elution and amplification of selected clones, with four to five cycles of selection usually being required. Cell bound antigens can be either found naturally on the cell surface or be there as a result of cloning and expression experiments.

Theoretically the affinity of the antibodies obtained is related to the size of the starting library.[59] This has also been shown to be true experimentally, where the affinity of antibodies isolated from a 10^7 element subset of a far larger library were shown to be lower than the affinity of the antibodies derived from the large library.[49] Ways in which the affinity of antibodies derived from phage display libraries can be increased are discussed in the next section.

Methods to Select Phage Antibodies

The simplest way to select phage antibodies is to coat a plastic tube with the antigen, add a phage antibody library and allow it to interact with the antigen on the surface of the plastic, wash the tube thoroughly, then elute the phage antibodies which have bound. The bound phage can be amplified and the cycle of binding and elution can be repeated a number of times.[32,60] This method requires rather large amounts of pure antigen. Antigen bound columns can be used in a similar fashion. Selections can be performed with smaller quantities if the antigen is biotinylated prior to interaction with the phage antibody library. Binding phage are then rescued using avidin coupled to magnetic beads.[61] By varying the concentration of biotinylated antigen one can select phage antibodies with different properties.[61]

As an alternative to producing and purifying recombinant proteins for selection, the protein products of cloned genes can be displayed on the surface of bacteria;[62] such recombinant bacteria can be used as living columns to select antibodies,[63] a procedure which appears to be more efficient than using purified antigen and which should allow one to go from gene to antibody in a relatively simple step.

In the same way that interesting monoclonal antibodies have been obtained by immunizing mice with cells and characterizing the hybridomas which arise, so new and interesting antibody specificities can also be obtained by screening naïve (i.e., unimmunized) natural or synthetic phage antibody libraries against whole cells or cellular extracts attached to solid supports and characterizing the antibodies which are isolated.[64,65] The principle for selection in these cases is identical to those described above (repeated cycles of binding, washing and elution), except when whole cells are used phage antibodies can be rescued by lysing the cells. A different technique employing whole cells has been used to isolate phage antibodies recognizing different B cell subtypes: the B lymphocytes were interacted with a phage antibody library and then phage which bound to a specific cell type were isolated by FACS.[66]

In vivo methods of selection involving the reconstitution of phage infectivity by antigen antibody interaction are also being developed[67,68] but are not yet generally applicable. In principle, p3 is divided into two parts. The N terminal part has the antigen of interest fused to its C terminus, while the C terminal part of p3 has an antibody fused to its N terminus and is attached to the phage in the normal fashion via the C terminal hydrophobic tail. Phage containing antibody / truncated p3 are unable to infect unless they bind an antigen attached to the missing part of p3, recreating a complete and functional p3 (see Fig. 3.6). This has been shown to work in a model system, but has not yet been applied to purify phage antibodies against recombinant antigens from a library.

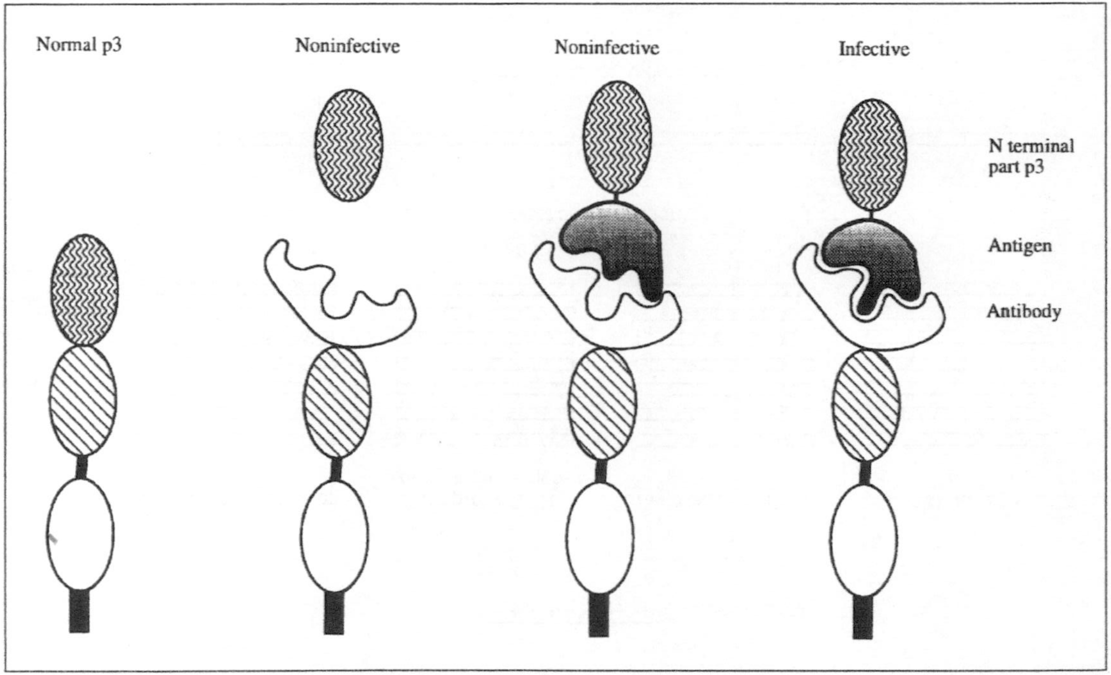

Fig. 3.6. Selectively infective phage. A novel experimental way to select phage antibodies which couples selection to infection is illustrated. p3 is divided between the penetration and receptor domains. An antigen is either expressed or coupled to the penetration domain and an antibody to the rest of p3 including the receptor domain. Infection is only possible when the two parts are brought close together as a result of an antigen/antibody interaction.

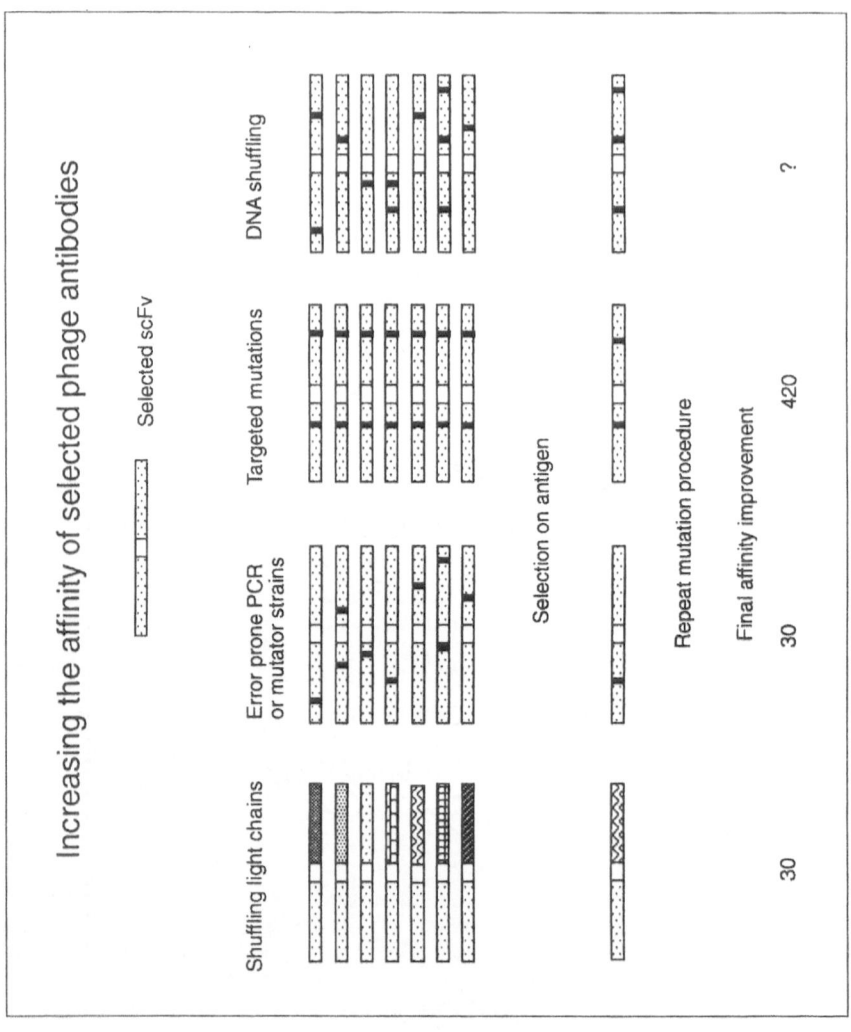

Fig. 3.7. Improvement of the affinity of phage antibodies. All methods used to improve the affinity of phage antibodies rely upon the creation of sub-libraries of mutated antibodies based on the original phage antibody. Such libraries are then used as the basis for further cycles of selection. The sub-libraries are either created in a targeted fashion (i.e., mutations are targeted to the CDRs) or in a random fashion using error prone PCR, DNA shuffling or mutator strains. A different sub-library is created by fixing the heavy chain and shuffling the light chains, this being repeated for the heavy chain when an improved antibody has been selected. The final reported affinity improvement is indicated.

Improvement of Antibody Affinity

The affinity of antibodies isolated from phage display libraries can be increased by a number of different mechanisms (see Fig. 3.7):

    a. random mutations using mutator strains;[69]
    b. targeted mutations;[70-72]
    c. chain shuffling;[72,73,74]
    d. error prone PCR;[52,61,75]
    e. sexual PCR / DNA shuffling.[76]

When a selected phage antibody is expressed in a mutator strain, random mutations at the rate of approximately 1/2000 bp are inserted into the DNA. Some of these give rise to changes in antibody structure and hence affinity for the antigen under study. Passage of one or more specific phage antibodies through a number of cycles of mutation in a mutator strain has been shown to increase the affinity of the derived phage antibodies by 100-fold.[69] This is a method which requires very little manipulation except for the passage from one strain to another and the harvesting of mutated phage.

An alternative to the use of mutator strains is the use of error prone PCR. Nonproofreading enzymes used in PCR (e.g., Taq polymerase) have a natural error rate of 1/9000.[77] This can be increased by altering the PCR conditions (e.g., increasing Mg or nucleotide concentrations). If selected V regions are reamplified using error prone PCR, a library of derivative V regions will be obtained from which new binding specificities can be selected after redisplay on phage. Error prone PCR and chemical mutagenesis used in this way has increased the affinity of an anti carbohydrate ScFv by 10-fold.[75] This method has the advantage of targeting the mutations to the V region only, whereas mutator strains produce mutations in the whole plasmid.

An alternative to the use of random mutations is the creation of small sublibraries derived from selected antibodies in which specific sites known or suspected to be involved in antigen binding (on the basis of structural modeling) have been randomized by PCR. Mutations isolated in different libraries (where different amino acids are mutated) can often be combined to produce an additive increase in affinity. By combining some of the best mutations of five different libraries, this method was used to increase the affinity of antibodies recognizing the HIV gp120 protein by up to 420-fold, reaching a final affinity of 15 pM.[70]

When using error prone PCR or mutation strains, mutations are strictly heirarchical, that is mutations occurring in separate molecules cannot be combined, unless they occur again independently; an event which is extremely unlikely. A recently developed mutation method based on PCR, termed DNA shuffling or sexual PCR,[78] allows mutations in separate molecules to be combined at random. In this method PCR amplified DNA is digested with DNAse (to small fragments) and reassembled using PCR. Mutations which were on different molecules now have the opportunity to come together (see Fig. 3.7). This method has been applied to antibodies, but under the experimental system used (ScFv display on phagemid),[76] probably induced a shortening of the ScFv linker with a subsequent dimerizaton or higher order oligomerizaton of the ScFv. This caused an apparent increase in affinity of the phage ScFv which was not seen in the soluble ScFv. Such a method is likely to be effective if the nature of the linker is changed, so that it is not repetitive or, alternatively, if Fab fragments are used.

Instead of mutating V regions to increase their affinity, a procedure called chain shuffling[79] can be used. In this technique the heavy chain V region from the best phage antibody isolated is kept constant and displayed in association with a library of light chains, the procedure then being reversed with a library of heavy chain V regions displayed with the best previously obtained light chain V region (or vice versa). In this way, antibodies with affinities up to 300 times greater than the initial selected phage antibody have been derived.[79,80]

By using a combination of many of the above mentioned methods, a total improvement of 1200-fold in the affinity of an anti erbB-2 ScFv fragment was achieved, yielding to the isolation of an improved ScFv with picomolar affinity.[81]

Chain shuffling has been shown to result in a slight change of antigen specificity in some cases.[74,82] This is not also true for the other methods has not been conclusively demonstrated.

Choice of Vector System: Phage or Phagemid

Two vector systems have been used to display antibodies on the surface of filamentous phage. The first and theoretically most straightforward is to directly modify the phage DNA itself. In this case all phage will carry three to five copies of the chimeric p3 as well as the DNA which encodes it. The second is to use phagemids. These are plasmids which contain an Ff origin of replication and so will be packaged into phage particles, when present in a bacteria in which all the other phage proteins are present. This is done by infecting bacteria containing a phagemid with a helper phage. The origin of replication of the helper phage is interrupted in such a way that phagemid DNA is packaged 10-50 times more efficiently than helper phage DNA. When used for phage display, the phagemid also contains a copy of the chimeric gene 3, under the control of a promoter, such as lac. This allows the production of a large excess of chimeric p3 protein, incorporated into the phage as well as the normal p3 provided by the helper phage. Phagemids have the advantage that the DNA is far easier to manipulate than that of phage, but the disadvantage that, notwithstanding the high level of p3 production, less than 10% of phagemids contain chimeric p3.[83] This can be an advantage in the isolation of high affinity antibodies, as avidity effects can be avoided.

Choosing the V Region Form

Antibody engineering has provided a wealth of different molecules based on the basic antibody structure (see Fig. 3.8). The full length immunoglobulin has the antigen binding VH and VL domains attached to constant regions, containing three domains in the case of IgG heavy chain and one domain in the case of the light chain. Originally defined by cleavage with the protease papain, Fab fragments contain a variable and constant domain for both heavy and light chain. These are now far more easily made using recombinant techniques. The Fv fragment consists of the VH and VL domains held together by noncovalent forces. These are usually not strong enough to ensure that the heavy and light chain are always associated. This problem is somewhat overcome in the ScFv fragment, in which heavy and light chain V regions are connected by a flexible linker peptide. These linker peptides are generally composed of amino acids usually found in flexible protein segments (glycine, serine, threonine),[16] although they have also been derived from naturally occurring linkers.[84] If the linker peptide is shorter than fifteen amino acids (usually from 0 to 5 amino acid residues) the VH and VL are unable to associate with

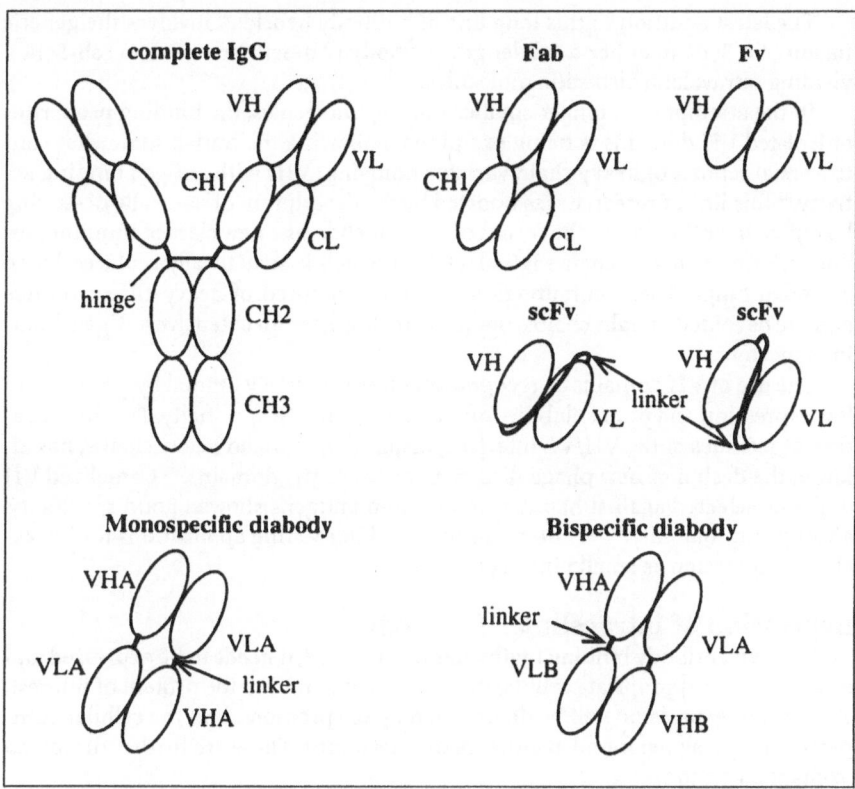

Fig. 3.8. The structure of antibodies and fragments derived from them. A complete antibody comprises two heavy chains and two light chains with the two antigen binding regions made up of a heavy and light chain variable region. The VH-CH1/VL-CL molecule is termed a Fab and is monomeric in its binding ability. Monomeric binding units can also be created by expressing the variable regions, either alone (Fv) or covalently joined by a linker (scFv - single chain Fv). It has been found that scFvs have the tendency to dimerize. This has been exploited in the creation of either monospecific or bispecific 'diabodies'.

one another and a dimeric ScFv form (termed a diabody) is formed.[85-87] This can be made either monospecific if a single pair of VH and VL domains joined by a single linker peptide are expressed, or dimeric if two ScFv domains containing nonhomogenous VH and VL domains are simultaneously expressed (in the order VHA-VLB and VHB-VLA, or VLA-VHB and VLB-VHA). Other methods for the creation of bivalent and/or bispecific (or even tetravalent) have also been described, which involve the use of dimerizaton[88] or tetramerizaton domains[89] (e.g., based on amphipathic coiled coil helix).

The use of the chelate effect to make high-affinity antibodies was exploited by creating a bispecific antibody fragment (Chelating Recombinant Antibody or CRAb)[90] that recognizes adjacent and nonoverlapping epitopes of the target antigen, and is flexible enough (through a peptide linker connecting the two ScFv fragments) to bind to both epitopes simultaneously. The CRAb was shown to have a much higher affinity than either of the ScFv fragments.

The latest addition in this long line of antibody hybrids[91] involves the genetic fusion of a ScFv to either a full length antibody (Ab-ScFv) or a Fab2 (Fab-ScFv), yielding tetravalent, bispecific molecules.

In the attempt to engineer smaller binding molecules, the binding properties of isolated VH domains is being examined. Following the initial, somewhat controversial reports of heavy chain variable domains (VH) with antigen binding activity,[92] this line of research was boosted by the description of naturally occurring heavy chain antibodies in the serum of the camelids, as a new class of immunoglobulins.[93] Camel heavy-chains IgGs lack CH1 which is structurally replaced by an extended hinge. These immunoglobulins are composed of heavy-chains dimers and are devoided of light chains, but nevertheless have an extensive antigen-binding repertoire.

The use of VH domains as recognition units is severely limited by problems of low expression and poor solubility often resulting in low specificity. The modification of residues at the VH/VL interface, inspired by camelid heavy chains, has allowed the design of new phage display libraries of VH domains.[94] Camelized VH domains selected against haptens and protein antigens showed good specificity, affinity and solubility.[95] VH domains may find interesting applications for intracellular expression as small binding units.

Expression of Intracellular Antibodies

Once an antibody binding region has been cloned, it needs to be expressed and directed to the appropriate cellular compartment to inhibit the protein of interest. The level of expression will be determined by the promoter and the cellular compartment by a signal found at either N or C terminus. These are further discussed in chapters 5 and 7.

An integrated vector system designed for the expression of ScFv fragments in different intracellular compartments after selection by phage display or cloning from hybridomas has been described.[30,96] This uses restriction sites for the cloning of V regions which are rare in human V regions, and has vector elements separated by restriction enzyme sites in such a way that promoter and selection marker can be easily modified. Vectors targeting to cytoplasm, nucleus, mitochondria, endoplasmic reticulum and secretory pathway were constructed; the modular nature of the vector permits the easy creation of further new vectors targeting to other compartments or carrying novel effector functions.

References

1. Kabat EA, Wu TT, Reid-Miller M et al. Sequences of proteins of immunological interest. US Department of Health and Human Services, US Government Printing Office 1987.
2. Chothia C, Lesk A. The relation between the divergence of sequence and structure in proteins EMBO J 1986; 5:823-826.
3. Chothia C, Lesk A. Canonical structures for the hypervariable regions of immunoglobulins. J. Biol Chem. 1987; 196:901-917.
4. Tramontano A, Chothia C, Lesk A. Framework residue 71 is a major determinant of the position and conformation of the second variable region in the VH domains of immunoglobulins. J Mol Biol 1990; 215:175-182.
5. Wilson IA, Stanfield RL. Antibody-antigen interactions. Curr Opin Struct Biol 1994; 3:113-118.

6. MacCallum RM, Martin ACR, Thornton JM. Antibody-antigen interactions: contact analysis and binding site topography. J Mol Biol 1996; 262:732-745.

7. Tonegawa S. Somatic generation of antibody diversity. Nature 1983; 302:575-581.

8. Milstein C. The Croonian Lecture 1989; Antibodies: a paradigm for the biology of molecular recognition. Proc R Soc London Biol. 1990; 239:1-16.

9. Neuberger MS, Milstein C. Somatic hypermutation. Curr Opin Immunol 1995; 7:248-254.

10. Tomlinson IM, Walter G, Jones PT et al. The imprint of somatic hypermutation on the repertoire of human germline V genes. J Mol Biol 1996; 256:813-817.

11. Neri D, Montigiani S, Kirkham PM. Biophysical methods for the determination of antibody-antigen affinities. Trends in Biotechnology 1996; 14:465-470.

12. Crothers DM, Metzger H. The influence of polyvalency on the binding properties of antibodies. Immunochem 1972; 9:341-357.

13. Bradbury A, Ruberti F, Werge T et al. The cloning of hybridoma V regions for their ectopic expression in intracellular and intercellular immunizaton. In: Borrebaeck C, ed. Antibody Engineering II: A Practical Approach. New York: IRL Press, 1994; 295-361.

14. Kabat EA, Wu TT, Perry HM. et al. Sequences of Proteins of Immunological Interest, 5 ed. U.S. Department of Health and Human Services, U.S. Government Printing Office, 1991.

15. Ruberti F, Cattaneo A, Bradbury A. The use of the RACE method to clone hybridoma cDNA when V region primers fail. J Imm Methods 1994; 173:33-39.

16. Bird RE, Hardman KD, Jacobson JW et al. Single-chain antigen-binding proteins. Science 1988; 242:423-426. [published erratum appears in Science 1989 Apr. 28;244(4903):409].

17. Bird RE, Walker BW. Single chain antibody variable regions. Trends Biotech 1991; 9:132-138.

18. Orlandi R, Gussow DH, Jones PT et al. Cloning immunoglobulin variable domains for expression by the polymerase chain reaction. Proc Natl Acad Sci USA 1989; 86:3833-3837.

19. Jones ST, Bendig M. Rapid PCR-cloning of full-length mouse immunoglobulin variable regions. Bio/Technology 1991; 9:88-89.

20. Orum H, Andersen PS, Oster A et al. Efficient method for constructing comprehensive murine Fab antibody libraries displayed on phage. Nucleic Acids Res 1993; 21:4491-8.

21. Sastry L, Alting MM, Huse WD et al. Cloning of the immunological repertoire in Escherichia coli for generation of monoclonal catalytic antibodies: construction of a heavy chain variable region-specific cDNA library. Proc Natl Acad Sci USA 1989; 86:5728-32.

22. Chaudhary V, Batra JK, Gallo MG et al. A rapid method of cloning functional variable-region antibody genes in Escherichia coli as single-chain immunotoxins [published erratum appears in Proc Natl Acad Sci USA 1990; Apr;87(8):3253]. Proc Natl Acad Sci USA 1990; 87:1066-70.

23. Winter G. Primers for amplifying mouse V regions; EMBO course on making antibodies in bacteria and on phage, 1991.

24. Larrick JW, Danielsson L, Brenner CA et al. Rapid cloning of rearranged immunoglobulin genes from human hybridoma cells using mixed primers and the polymerase chain reaction. Biochem Biophys Res Commun 1989; 160:1250-1256.

25. Larrick JW, Danielsson L, Brenner CA et al. Polymerase chain reaction using mixed primers: Cloning of human monoclonal antibody variable region genes from single hybridoma cells. Bio/Technology 1989; 7:934-938.

26. Campbell MJ, Zelenetz AD, Levy S et al. Use of family specific leader region primers for PCR amplification of the human heavy chain variable region gene repertoire. Mol Immunol. 1992; 29:193.

27. Ostermeier C, Michel H. Improved cloning of antibody variable regions from hybridomas by an antisense-directed RNase H digestion of the P3-X63-Ag8.653 derived pseudogene mRNA. Nuc Acids Res 1996; 24:1979-1980.

28. Frohman MA, Dush MK, Martin G. Rapid production of full length cDNAs from rare transcripts: Amplification using a single gene-specific oligonucletide primer. Proc Natl Acad Sci USA 1988; 85:8998-9002.

29. Edwards JB, Delort J, Mallet J. Oligodeoxyribonucleotide ligation to single stranded cDNAs: a new tool for cloning 5' ends of mRNAs and for constructing cDNA libraries by in vitro amplification. Nuc Acids Res 1991; 19:5227-5232.

30. Persic L, Roberts A, Wilton J et al. An integrated vector system for the eukaryotic expression of antibodies or their fragments after selection from phage display libraries. Gene 1997; 187:9-18.

31. Heinrichs A, Milstein C, Gherardi E. Universal cloning and direct sequencing of rearranged antibody V genes using C region primers, biotin-captured cDNA and one-side PCR. J. Immunol. Methods 1995; 178:241-251.

32. Nissim A, Hoogenboom HR, Tomlinson IM et al. Antibody fragments from a 'single pot' phage display library as immunochemical reagents. EMBO J 1994; 13:692-698.

33. Anand NN, Mandal S, Mackenzie CR et al. Bacterial expression and secretion of varius single chain Fv genes encoding proteins specific for a Samonella Serotype B O antigen. J Biol Chem 1991; 266:21874-21879.

34. Hoogenboom HR, Griffiths AD, Johnson KS et al. Multi-subunit proteins on the surface of filamentous phage: methodologies for displaying antibody (Fab) heavy and light chains. Nucl Acids Res 1991; 19:4133-4137.

35. Cabilly S. Growth at sub-optimal temperatures allows the production of functional, antigen-binding Fab fragments in Escherichia coli. Gene 1989; 85:553-557.

36. Field H, Yarranton GT, Rees AR. Expression of mouse immunoglobulin light and heavy chain variable regions in Escherichia coli and reconstitution of antigen-binding activity. Protein Eng 1990; 3:641-7.

37. Buchner J, Brinkmann U, Pastan I. Renaturation of a single chain immunotoxin facilitated by chaperones and protein disulfide isomerase. Bio/Technology 1992; 10:682-685.

38. Walls MA, Hsiao K, Harris LJ. Vectors for the expression of PCR-amplified immunoglobulin variable domains with human constant regions. Nucl Acids Res 1993; 21:2921-2929.

39. Trill JJ, Shatzman AR, Ganguly S. Production of monoclonal antibodies in COS and CHO cells. Curr Opin Biotech 1995; 6:553-560.

40. Bauer M, Smith GP. Filamentous phage morphogenetic signal sequence and orientation of DNA in the virion and gene-V protein complex. Virology 1988; 167:166-75.

41. Bross P, Bussmann K, Keppner W et al. Functional analysis of the adsorption protein of two filamentous phages with different host specificities. J Gen Microbiol 1988; 134:461-471.

42. Stengele I, Bross P, Garces X et al. Dissection of functional domains in phage fd adsorption protein. Discrimination between attachment and penetration sites. J Mol Biol 1990; 212:143-9.

43. Smith GP. Filamentous fusion phage: novel expression vectors that display cloned antigens on the virion surface. Science 1985; 228:1315-1317.

44. Bradbury A, Cattaneo A. The use of phage display in neurobiology. TINS 1995; 18:243-249.

45. McCafferty J, Griffiths AD, Winter G et al. Phage antibodies: filamentous phage displaying antibody variable domains. Nature 1990; 348:552-4.
46. Barbas CF, Kang AS, Lerner RA et al. Assembly of combinatorial antibody libraries on phage surfaces: The gene III site. Proc Natl Acad Sci USA 1991; 88:7978-7982.
47. Breitling SD, Seehaus T, Klewinghaus I et al. A surface expression vector for antibody screening. Gene 1991; 104:147-153.
48. Kang AS, Barbas CF, Janda KD et al. Linkage of recognition and replication functions by assembling combinatorial antibody Fab libraries along phage surfaces. Proc Natl Acad Sci USA 1991; 88:4363-6.
49. Griffiths AD, Williams SC, Hartley O et al. Isolation of high affinity human antibodies directly from large synthetic repertoires. EMBO J 1994; 13:3245-3260.
50. Griffin HM, Ouwehand WH. A human monoclonal antibody specific for the leucine-33 (P1A1, HPA-1a) form of platelet glycoprotein IIIa from a V gene phage display library. Blood 1995; 86:4430-4436.
51. Ames RS, Tornetta MA, Jones CS et al. Isolation of neutralizing anti-C5a monoclonal antibodies from a filamentous phage monovalent Fab display library. J Immunol 1994; 152:4572-81.
52. Gram H, Marconi L, Barbas CF et al. In vitro selection and affinity maturation of antibodies from a naive combinatorial immunoglobulin library. Proc Natl Acad Sci USA 1992; 89:3576-3580.
53. Davies EL, Smith JS, Birkett CR et al. Selection of specific phage-display antibodies using libraries derived from chicken immunoglobulin genes. J Immunol Methods 1995; 186:125-135.
54. Zebedee SL, Barbas CF, Hom Y et al. Human combinatorial antibody libraries to hepatitis B surface antigen. Proc Natl Acad Sci USA. 1992; 89:3175-3179.
55. Barbas CF, Collet TA, Amberg W et al. Molecular Profile of an Antibody Response to HIV-1 as Probed by Combinatorial Libraries. J Mol Biol 1993; 230:812-823.
56. Cai X, Garen A. Anti-melanoma antibodies from melanoma patients immunized with genetically modified autologous tumor cells: selection of specific antibodies from single-chain Fv fusion phage libraries. Proc Natl Acad Sci USA 1995; 92:6537-6541.
57. Marks JD, Hoogenboom HR, Bonnert TP et al. By-passing immunization—human antibodies from V-gene libraries displayed on phage. J Mol Biol 1991; 222:581-597.
58. Hoogenboom HR, Winter G. Bypassing immunizaton: human antibodies from synthetic repertoires of germ line VH-gene segments rearranged in vitro. J Mol Biol 1992; 227:381-388.
59. Perelson AS, Oster GF. Theoretical studies of clonal selection: Minimal antibody repertoire size and reliability of self nonself discrimination. J Theor Biol 1979; 81:645-670.
60. Hoogenboom HR, Raus JCM, Volckaert G. Cloning and expresion of a chimeric antibody directed against the human transferrin receptor. J Immunol 1990; 144:33211-3217.
61. Hawkins RE, Russell SJ, Winter G. Selection of phage antibodies by binding affinity: mimicking affinity maturation. J Mol Biol 1992; 226:889-896.
62. Hofnung M. Expression of foreign polypeptides at the Escherichia coli cell surface. Methods Cell Biol 1991; 34:77-105.
63. Bradbury A, Persic L, Werge T, Cattaneo A. From gene to antibody: the use of living columns to select specific phage antibodies. BioTech 1993; 11:1565-1569.
64. Marks JD, Ouwehand WH, Bye JM et al. Human antibody fragments specific for human blood group antigens from a phage display library. Biotechnology (NY) 1993; 11:1145-9.

65. Merz DC, Dunn RJ, Drapeau P. Generating a phage display antibody library against an identified neuron. J Neurosci Meth 1995; 62:213-219.

66. de Kruif J, Terstappen L, Boel E et al. Rapid selection of cell subpopulation-specific human monoclonal antibodies from a synthetic phage antibody library. Proc Natl Acad Sci USA 1995; 92:3938-3942.

67. Krebber C, Spada S, Desplancq D et al. Coselection of cognate antibody-antigen pairs by selectively-infective phages. FEBS Lett 1995; 377:227-231.

68. Duenas M, Malmborg AC, Casalvilla R et al. Selection of phage displayed antibodies based on kinetic constants. Mol Immunol 1996; 33:279-285.

69. Low NM, Holliger PH, Winter G. Mimicking somatic hypermutation: affinity maturation of antibodies displayed on bacteriophage using a bacterial mutator strain. J Mol Biol 1996; 260:359-368.

70. Yang WP, Green K, Pinz-Sweene S et al. CDR walking mutagenesis for the affinity maturation of a potent human anti-HIV-1 antibody into the picomolar range. J Mol Biol 1995; 254:392-403.

71. Thompson J, Pope T, Tung JS et al. Affinity maturation of a high-affinity human monoclonal antibody against the third hypervariable loop of human immunodeficiency virus: use of phage display to improve affinity and broaden strain reactivity. J Mol Biol 1996; 256:77-88.

72. Schier R, Balint RF, McCall A et al. Identification of functional and structural amino acid residues by parsimonious mutagenesis. Gene 1996; 169:147-155.

73. Marks JD, Griffiths AD, Malmqvist M et al. By-passing immunization: building high affinity human antibodies by chain shuffling. Biotechnology (NY) 1992; 10:779-83.

74. Ohlin M, Owman H, Mach M et al. Light chain shuffling of a high affinity antibody results in a drift in epitope recognition. Mol Immunol 1996; 33:47-56.

75. Deng SJ, MacKenzie CR, Sadowska J et al. Selection of antibody single-chain variable fragments with improved carbohydrate binding by phage display. J Biol Chem 1994; 269:9533-8.

76. Crameri A, Cwirla S, Stemmer WP. Construction and evolution of antibody-phage libraries by DNA shuffling. Nature Med 1996; 2:100-102.

77. Tindall KR, Kunkel TA. Fidelity of DNA synthesis by the Thermus aquaticus DNA polymerase. Biochemistry 1988; 27:6008-6013.

78. Stemmer WP. Rapid evolution of a protein in vitro by DNA shuffling [see comments]. Nature 1994; 370:389-391.

79. Marks JD, Griffiths AD, Malmqvist M et al. By-passing immunization: building high affinity human antibodies by chain shuffling. BioTechnology 1992; 10:779-783.

80. Schier R, Bye J, Apell G et al. Isolation of high-affinity monomeric human anti-c-erbB-2 single chain Fv using affinity-driven selection. J Mol Biol 1996; 255:28-43.

81. Schier R, McCall A, Adams GP et al. Isolation of picomolar affinity anti-c-erbB-2 Single-chain Fv by molecular evolution of the complementarity determining regions in the center of the antibody binding site. J. Mol. Biol. 1996; in press.

82. Watkins BA, Davis AE, Fiorentini S et al. Evidence for distinct contributions of heavy and light chains to restriction of antibody recognition of the HIV-1 principal neutralization determinant. J Immunol 1996; 156:1676-1683.

83. Clackson T, Wells JA. In vitro selection from protein and peptide libraries. Trends in Biotech 1994; 12:173-184.

84. Takkinen K, Laukkanen MJ, Sizmann D et al. An active single-chain antibody containing a cellulase linker domain is secreted by *Escherichia coli*. Prot Eng 1991; 4:837-841.

85. Holliger P, Prospero T, Winter G. "Diabodies": small bivalent and bispecific antibody fragments. Proc Natl Acad Sci USA 1993; 90:6444-6448.
86. Desplancq D, King DJ, Lawson AD et al. Multimerization behavior of single chain Fv variants for the tumour-binding antibody B72.3. Prot Eng 1994; 7:1027-1033.
87. Perisic O, Webb PA, Holliger P et al. Crystal structure of a diabody, a bivalent antibody fragment. Structure 1994; 2:1217-1226.
88. Pack P, Plueckthun A. Miniantibodies: use of amphipathic helices to produce functional, flexibly linked dimeric Fv fragments with high avidity in *E. coli*. Biochemistry 1992; 31:1579-1584.
89. Pack P, Mueller K, Zahn R et al. Tetravalent miniantibodies with high avidity assembling in *E. coli*. J Mol Biol 1995; 246:28-34.
90. Neri D, Momo M, Prospero T et al. High-affinity antigen binding by chelating recombinant antibodies (CRAbs). J Mol Biol 1995; 246:367-373.
91. Coloma MJ, Morrison SL. Design and production of novel tetravalent bispecific antibodies. Nature Biotechnology 1997; 15:159-163.
92. Ward ES, Gussow D, Griffiths AD et al. Binding activities of a repertoire of single immunoglobulin variable domains secreted from Escherichia coli Nature 1989; 341:(6242):544-6.
93. Hamers-Casterman C, Atarhouch T, Muyldermans S et al. Naturally occurring antibodies devoid of light chains Nature 1993; 363:446-448.
94. Davies J, Riechmann L. Antibody VH domains as small recognition units. BioTechnology 1995; 13:475-479.
95. Davies J, Riechmann L. Single antibody domains as small recognition units: design and in vitro antigen selection of camelized, human VH domains with improved protein stability. Prot Engin 1996; 9:531-537.
96. Persic L, Righi M, Roberts A et al. Targeting vectors for intracellular immunization. Gene 1997; 187:1-8.

CHAPTER 4

Assembly and Folding of Antibodies in Natural and Artificial Environments

Antonino Cattaneo and Silvia Biocca

The biosynthesis of antibodies of the different classes, in lymphoid cells of the B lineage is a highly regulated process involving many regulatory events at the transcriptional, translational and post-translational level. Many of these regulatory events are lymphoid specific and can collectively be seen as part of a " quality control system" for antibody synthesis, oligomerization and expression. For proteins destined to the extracellular space, the term quality control defines the phenomena by which—in general—only properly folded and assembled molecules are secreted or expressed on the cell surface. Most aspects of these post-translational regulatory processes involve interactions with the constant antibody domains, occurring in the secretory pathway of B lymphocytes and plasma cells. Given the wide range of applications that antibodies have in science and technology, learning how this is achieved in cells of the B lineage is of considerable interest from the practical point of view.

As shall be seen antibodies can be expressed and secreted from nonlymphoid cells, but part of this tight "quality control" regulation is lost because of the different intracellular context. When it comes to expressing recombinant forms of antibodies lacking altogether, the constant domains such as ScFv fragments, their properties in the secretory pathway become crucially and unpredictably dependent on the particular variable region being expressed. This finds a parallel in the great variability of expression levels and in solubility of ScFv fragments expressed in *E. coli*. Recent studies have started to shed some light on ways to engineer antibody variable regions to improve their folding stability and solubility. In the end, these studies will be of great relevance to improve the ectopic expression of antibodies in different cellular contexts.

In this chapter we shall discuss some recent findings on the assembly and folding of Igs in their normal environment, the secretory pathway of lymphoid cells, as well as that of recombinant antibody forms in the more artificial environment of bacterial cells.

Intracellular Antibodies: Development and Applications, edited by Antonino Cattaneo and Silvia Biocca. © Springer-Verlag and Landes Bioscience 1997.

The Natural Environment:
The Secretory Pathway of B Lymphoid Cells

Quality Control of Newly Synthesized Proteins

The complex problem of protein folding is further complicated for multimeric proteins, such as antibodies, which derive from the correct assembly of single sub-units. Although the three-dimensional structure of proteins is dictated by their amino acid sequences,[1] it has become clear that folding, assembly and oligomer-ization are catalyzed in vivo by a family of 'chaperone' molecules;[2] moreover, these processes are tightly linked to 'quality control' events which ensure that only pro-teins that attain the proper tertiary and quaternary structure reach their final des-tination within the cell. This is particularly evident in the case of secretory or mem-brane proteins, like Igs.

All antibody classes share the basic H_2L_2 structure, and can be produced in two forms during B cell development: as membrane proteins on the surface of B lym-phocytes, where they act as antigen receptors, or as secretory effector molecules released by plasma cells. The two forms are produced by alternate RNA processing.

Heavy chains destined for membrane insertion (H_μ) have a stretch of hydro-phobic amino acids in their C-terminal end, which allows their insertion in the lipid bilayer. In contrast, hydrophilic peptides are found at the C-terminal ends of secreted heavy chains.

Having a N-terminal leader sequence, H and L chains are cotranslationally trans-located in the endoplasmic reticulum (ER), where membrane and secretory pro-teins begin to fold, assemble and oligomerize.[3-6] For oligomeric proteins such as secretory IgM, control events are particularly important. The rich array of chaper-one molecules residing in this organelle is thought to play a major role in these processes. Molecules which fail to acquire their tertiary and quaternary structure are generally prevented from reaching the distal organelles of the secretory path-way: the ER is equipped not only with folding and retention machineries, but also with a proteolytic system, independent from that of lysosomes.[7-8]

What kind of signals allow cells to discriminate between assembled and unassembled molecules? How is retention coupled to degradation? Where does quality control take place?

Ig Assembly in B and Plasma Cells

Figure 4.1 summarizes our understanding of the folding, assembly and poly-merization pathways of newly synthesized Igs in cells of the B cell lineage.

Subunit folding generally precedes assembly, and starts cotranslationally, in-dependently for H and L, and proceeds vectorially, from the V to the C domains. Experimentally, folding can be monitored by the formation of the intradomain disulfide bond linking the two conserved cysteines: this bond is formed shortly after translation of the domain is completed.[9-10] The vectoriality of folding pre-vents aberrant intrachain disulfide bond formation and the backward retrotranslocation of the newly synthesized growing chains from the ER lumen to the cytosol.[11] However, folding (and chain assembly) does not need to occur co-translationally, since both intra- and interchain disulfide bonds can be formed in Igs after translation is completed,[12-14] a point of some relevance for the folding of antibody domains in ectopic environments. Like subunit folding, intermolecular

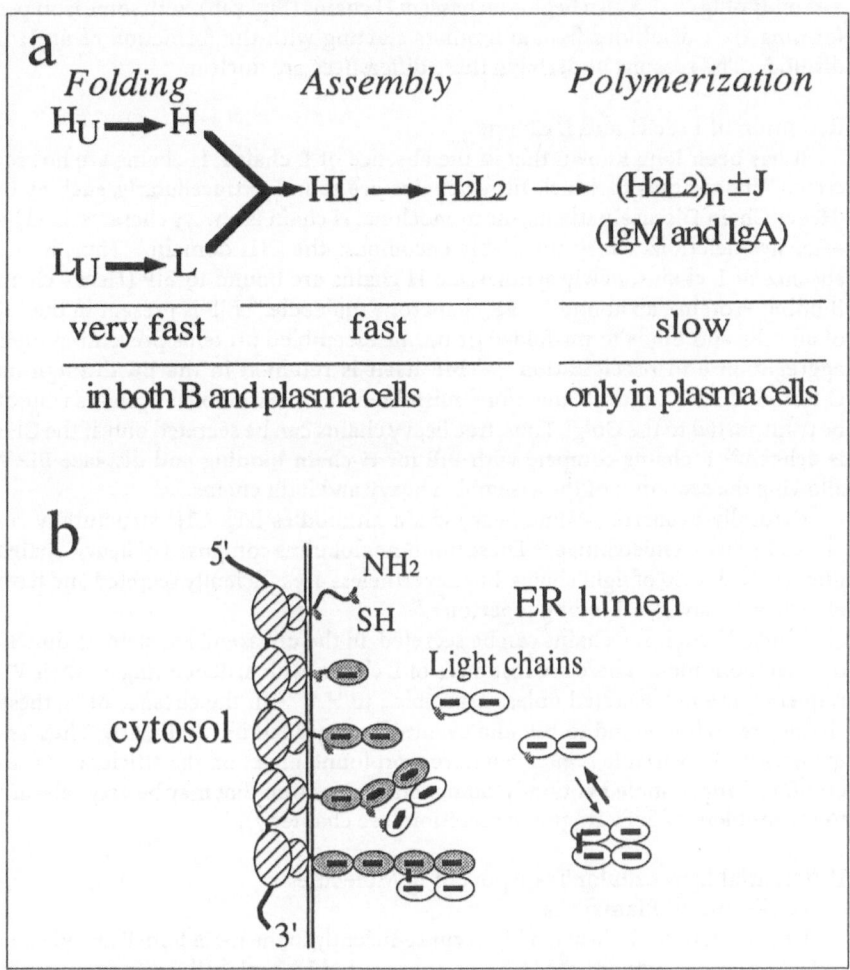

Fig. 4.1. Ig folding, assembly and polymerization occur sequentially in the ER.

(a) Rapid folding and assembly of newly synthesized Igs. In both B and plasma cells, unfolded H and L chains (Hu and Lu) rapidly fold and assemble into H_2L_2 "monomers" ($t^{1/2} \approx 2$ and 5 min respectively). Polymerization is slow ($t^{1/2} \geq 60$ min) and does not take place in B lymphocytes.

(b) Folding and assembly begin cotranslationally. The N-terminal domains begin to fold while the nascent H chain is still on the polysome. When at least the VH and CH1 have folded and the intradomain disulfides formed, nascent H may assemble with pre-existing L chains. Some L chains form L_2 homodimers. In B cells, Hm and Hs homodimerization might also involve chains synthetized on the same polysome. Reprinted with permission from ref. 82.

assembly of Ig chains also begins on nascent H chains (Fig. 4.1b), with some isotypes forming H-H disulfides first, and others starting with the formation of an H-L disulfide. The reasons underlying these differences are unclear.

Retention of Free H and L Chains

It has been long known that in the absence of L chains, H chains are not secreted. Those cases in which heavy chains are found extracellularly, such as in 'Heavy Chain Disease' patients, the monoclonal H chain is always characterized by extensive deletions which invariably encompass the CH1 domain.[15] Thus, in the absence of L chains, newly synthesized H chains are bound to BiP (Heavy chain Binding Protein), an abundant ER chaperone molecule.[16] BiP is present in the ER of all cells and binds to misfolded or un/misassembled proteins preventing their aggregation and precipitation.[17-21] BiP itself is retained in the ER through its C-terminal KDEL motif[22] therefore, misfolded or misassembled proteins cannot be transported to the Golgi. Thus, free heavy chains can be secreted only if the CH1 is deleted.[23] L chains compete with BiP for H chain binding and displace BiP,[24] allowing the secretion of the assembled heavy and light chains.

Naturally occurring camel heavy chain antibodies lack CH1 structurally replaced by an extended hinge.[25] These immunoglobulins composed of heavy-chains dimers are devoid of light chains, but nevertheless are efficiently secreted and have an extensive antigen-binding repertoire.[25]

Unlike H chains, L chains can be secreted in the unassembled state, as dimers or even monomers. There are instances of L chains which, depending on their VL sequence, are not secreted unless assembled to H.[26-29] In the absence of H, these chains are stably bound to BiP and eventually degraded intracellularly. Thus, sequences in the variable region can have a profound effect on the efficiency of secretion of unassembled antibody chains, an observation that may be very relevant to the problem of ScFv fragment secretion (see chapter 7).

Differential Intracellular Transport of Ig Molecules in B Cells and in Plamacells

Heavy and light chain assembly occurs efficiently in the ER of both B and plasma cells. As a result, monomeric H_2L_2 Igs are secreted by both cell types, in different amounts (plasma cells secreting much more efficiently) but essentially at similar rates. On the other hand, the expression of surface Ig, and the secretion of polymeric Igs (as IgM and IgA) are markedly different in B cells and in plasma cells.

Heavy chains destined for membrane insertion (H_μ) have a stretch of hydrophobic amino acids in their C-terminal end which allows their insertion in the lipid bilayer. The absence of tyrosine-kinase or other signaling activities in the cytoplasmic portions of $H\mu$ chains implies the association of the $H\mu_2L_2$ complex with other signaling molecules.[30]

IgM is expressed on the surface of virgin, immunocompetent B lymphocytes, while it is notably absent from the membrane of plasma cells. The absence of surface IgM in plasma cells is in part due to the preferential splicing of soluble H chain transcripts. When myeloma cell lines are transfected with constructs encoding only μm chains, the latter are efficiently assembled to L chains, but the μm_2L_2 complexes are arrested and degraded in the ER.[31-32] The failure to negotiate transport out of the ER is due to the absence of one of the accessory molecules necessary for assembly of a functional B cell receptor. Transport to the cell surface requires μm, L and two additional proteins, the products of the mb-1 and B29 genes (renamed α and β chains). Both α and β chains carry sequence motifs in their cytoplasmic portions that mediate signal transduction by recruiting downstream effectors (for a review see ref. 30 and 33).

Consistent with the need for these B cell specific accessory proteins to make a membrane IgM competent for transport to the cell surface, nonlymphoid cells as plasma cells, are unable to sustain the transport to the membrane of mIgM. Reconstitution experiments demonstrated conclusively that an antigen receptor can be reconstituted in nonlymphoid cells by coexpression of H and L chains together with the mb-1 and B29 gene products, with no other lymphoid-specific proteins needed.[33,49] Substitution of a small number of critical amino acids in the μ transmembrane segment to increase its hydrophobicity, relieves this intracellular retention and allows IgM to be transported to the plasma membrane in nonlymphoid cells as well.[49]

Opposite is the fate of secretory IgM. In fact, B cells retain and degrade intracellularly virtually all $μs_2L_2$ monomeric complexes. This is not due to a defective secretion in general, as B cells secrete other monomeric immunoglobulins,[34-35] such as IgG or IgE or mutant IgM in which the transmembrane portion of μm chains has been deleted.[36] Secretion of IgM is controlled at the level of assembly. The structure of secretory IgM is schematized in Figure 4.2. Two polymeric forms, either hexamers or pentamers containing a J chain, but not monomers, are secreted by plasma cells.[36-38] Thus, unlike IgG or other 'monomeric' Igs, monomeric IgM must further assemble into a polymer, in order to be transported through the Golgi and be secreted. The polymerization of IgM is due to the presence of the so-called 'secretory tailpiece' at the C-terminus of μ chains. Within the μs tailpiece, a conserved cysteine at the penultimate position forms the covalent bonds linking the H_2L_2 subunits into a polymer with or without J chains. Not only is this residue (Cys575) essential for IgM polymerization, but it is also responsible for the selective retention of unpolymerized μ2L2 subunits, in both B cells and plasma cells. Polymerization per se is not necessary for secretion; rather, it probably hinders the μs tailpiece, and Cys575 in particular. This residue is essential for IgM polymerization. Thus, in the case of IgM secretion, selectivity is achieved not by the expression of some 'positive' transport signal on structurally mature proteins, but by the masking of retention elements on assembly intermediates. This retention element, Cys575 in the μs tailpiece, mediates assembly, retention of unpolymerized intermediates and their targeting to degradation and acts as a three way switch, mediating assembly, retention and degradation of both IgM and cathepsin D chimeras (Fig. 4.3).[6,36]

These results reveal the hierarchy of events which couple the structural maturation of newly synthesized IgM to their intracellular transport. At least two retention elements are involved in the quality control of this isotype: first, the Cμ1 domain controls H-L assembly, second Cys575 is in charge of verifying that only polymers proceed to the Golgi. The logic of quality control is similar in both cases. It is assembly or polymerization which mask the retention element, allowing secretion. Mutant IgM lacking Cys575 are secreted also by B cell transfectants, indicating that this residue plays a similar role in preventing the secretion of unpolymerized IgM in both B and plasma cells. Thus, the problem of why B cells do not secrete IgM can be reduced to the question of why B cells are not able to form IgM polymers.

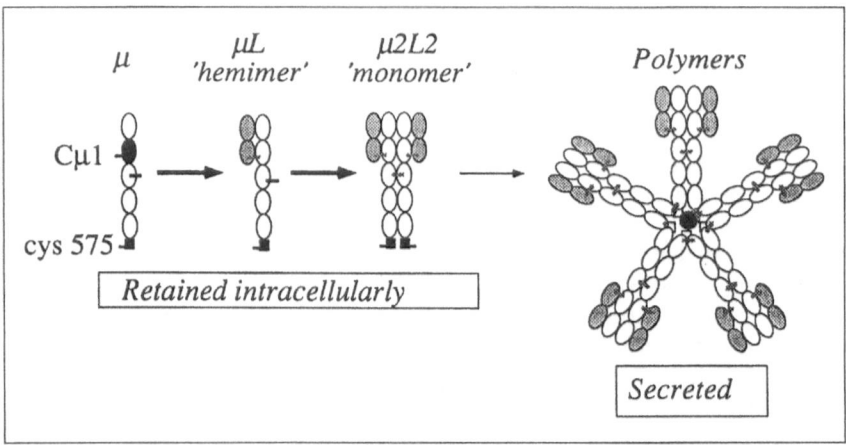

Fig. 4.2. Biogenesis of secretory IgM. At least two retention elements are present in secretory μ chains. In the absence of L chains, the Cμ1 (black) interacts with BiP. The carboxy terminal cysteine mediates retention of μ_2L_2 and other intermediates of polymerization. These retention elements are all masked in polymers. A pentamer with J chain (dark circle) is depicted in the figure. Note that hexamers devoid of J chain can also be secreted by both lymphoid and nonlymphoid cells.[42] Reprinted with permission from ref. 82.

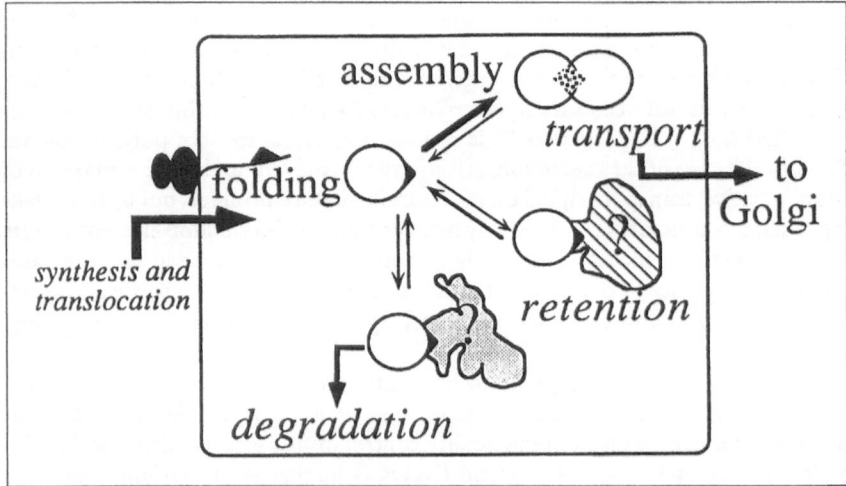

Fig. 4.3. Quality control in the ER. Cotranslational translocation into the ER, folding, assembly and transport to the Golgi are sequential events in the life of newly synthetized proteins. Unassembled subunits interact transiently with BiP, PDI and other resident proteins in the ER via the same element(s) that also mediate assembly and degradation (the dark triangle). Oligomerization will mask the signal and allow transport. In the scheme, degradation is mediated by the molecule drawn in grey, which might be either a receptor for degradation or a protease itself. Reprinted with permission from ref. 82.

It has long been thought that J chain is necessary for polymerization. Its synthesis correlate with the onset of IgM secretion.[40-41] However, B cells transfected with J chain, do not acquire a secretory phenotype, and nonlymphoid cells, which do not express J chain, secrete polymeric IgM when transfected with μ and L chain genes.[42] Thus, J chain is neither sufficient nor necessary for IgM secretion. Rather than controlling the secretion of polymeric Igs during lymphocyte development, J chain seems to determine the type of IgM polymer produced, pentamer versus hexamer.[42] In turn, this affects the effector properties of IgM.

The reasons underlying the failure of B cells to polymerize and hence to secrete IgM are presently unclear. In hybrids between B cells and plasma cells, the secretory phenotype of the latter is dominant, suggesting that either B cells lack some essential catalyst of IgM polymerization, or they express some specific inhibitor. Also *Xenopus laevis* oocytes—like B cells—fail to polymerize and secrete IgM (R. Sitia, personal communication): if a specific system exists in B cells which prevents IgM polymerization and secretion, it should be shared by *Xenopus* oocytes.

The role of disulfide interchange reactions in controlling IgM transport is emphasized by the observation that membrane permeant monovalent reducing agents (i.e., 2-mercaptoethanol [2ME] and N-acetyl cysteine) induce the secretion of unpolymerized Ig by both B and plasma cells,[6,43] while proteins which are retained intracellularly by mechanisms other than disulfide interchange are not secreted. A likely explanation is that the reducing agent competes with the recognition of the free thiol group involved in retention. In the ER, thiol mediated retention was suggested to involve the formation of reversible disulfide bonds with the protein matrix of the ER.[44]

The thiol-based quality control mechanisms, so efficient in the ER, becomes of little importance in or beyond the Golgi.[14] Once proteins have reached the Golgi, they can proceed through the distal sections of the secretory apparatus also in the unassembled state, suggesting that thiol-mediated retention is not operative in or beyond the Golgi.[14]

Thus, exposed thiols act as intracellular retention elements for unassembled or partially assembled secretory antibodies. Yet, some proteins can be secreted despite the presence of an unpaired cysteine. Among these are IgA and free monomeric L chains. How do these proteins escape thiol-mediated retention? A comparison of the amino acid sequences around the unpaired cysteine residues demonstrates that the amino acid context in which a given cysteine resides determines its efficacy as an ER retention element. In particular, the presence of an acidic residue next to the critical cysteine may allow the masking of the thiol and progress through the Golgi for both IgA[44] and light chains.[42]

These results show how a simple mechanism—the recognition of exposed thiols on assembly intermediates—can be precisely regulated to achieve a fine tuning of retention/degradation and secretion of immunoglobulins.

Expression of Antibodies in the Secretory Pathway of Nonlymphoid Cells

Following further the approach of comparing antibody assembly in B cells versus plasma cells, the natural and logical extension is to determine antibody assembly and secretion in nonlymphoid cells.

In the secretory pathway of nonlymphoid cells, by and large things are in common with their lymphoid counterpart. For instance, also in nonlymphoid cells the secretion of immunoglobulins follows a constitutive pathway.[42] The efficiency of secretion, however, varies from cell to cell.[42] In some nonlymphoid cell types the efficiency of secretion is remarkably efficient. In particular, this is the case for cells related to the nervous system. This forms the basis of the so called neuroantibody approach[42,46,47] (see chapter 6). Other common aspects include the fact that also nonlymphoid cells retain and degrade free heavy chains or unpolymerized IgM subunits with mechanisms similar to those acting in lymphoid hosts. Interestingly, nonlymphoid transfectants are able to secrete polymeric IgM even in the absence of the lymphoid specific J chain protein.[38,42] In contrast, plant cells are not able to secrete polymeric IgM, in the absence of J chain.[48]

The production of IgA has been compared in lymphoid and in nonlymphoid cells. Similarly to IgM, IgA are not secreted by B cells, but, unlike the IgM case, plasma cells can secrete both polymeric and monomeric IgA. Again, *Xenopus* oocytes behave like B cells, retaining and degrading IgA monomers intracellularly (R. Sitia, personal communication), while other nonlymphoid cells can secrete monomeric IgA.[53,54] IgA dimers are linked, in plasma cells, through a J chain. To reach external secretions, polymeric IgA must be transported across epithelial cells, which is mediated by the polymeric Ig receptor. The complex is internalized from the basolateral side, transported across the epithelial cell, from which it is released at the apical face. During transcellular transport, the receptor undergoes partial proteolysis and a fragment (the "secretory component" SC) becomes disulfide-bonded to dimeric IgA, to yield secretory IgA (SIgA). Thus, two distinct cell types are needed to assemble SIgA in mammals. This limits severely the efficiency with which secretory IgA molecules can be prepared for application purposes. In transgenic plants, equipped with the four genes necessary for making a SIgA, efficient assembly of SIgA can occur in single cells[55] (see chapter 10).

There are, however, some notable differences in the way in which the regulation of Ig biogenesis and transport occurs in lymphoid and in nonlymphoid cells. One example of this regulation has been discussed above and is provided by the plasma membrane transport of membrane IgM (see above).

Another example is provided by the secretion of Ig molecules tagged with the retention signal SEKDEL in myeloma cells. The SEKDEL-based retention system is ubiquitous, as it is found in all eukaryotic cells. This sequence, added at the C-terminus of antibody heavy or light chains and of ScFv fragments, leads to the retention of the corresponding polypeptide in the lumen of the ER, in different mammalian cell lines.[50,51] However, it was found that this signal fails to inhibit the secretion of whole antibody molecules from plasma cells.[52] Thus, these cells secrete SEKDEL-tagged assembled immunoglobulins as efficiently as the nontagged counterparts. This is not due to a proteolytic cleavage of the C-terminal signal, nor to the lack of the SEKDEL receptor in these cells, since light chains, that would normally be secreted, are efficiently retained by the SEKDEL tag. Rather, it appears that plasma cells are able to override the SEKDEL retention system, in order to achieve the secretion of an assembled antibody, whilst efficiently retaining an unassembled antibody chain. This also suggests that antibody secretion from plasma cells, rather than being a default process, as commonly thought, may involve the active participation of (unknown) signals or mechanisms, that are able to override a competing ER retention system.

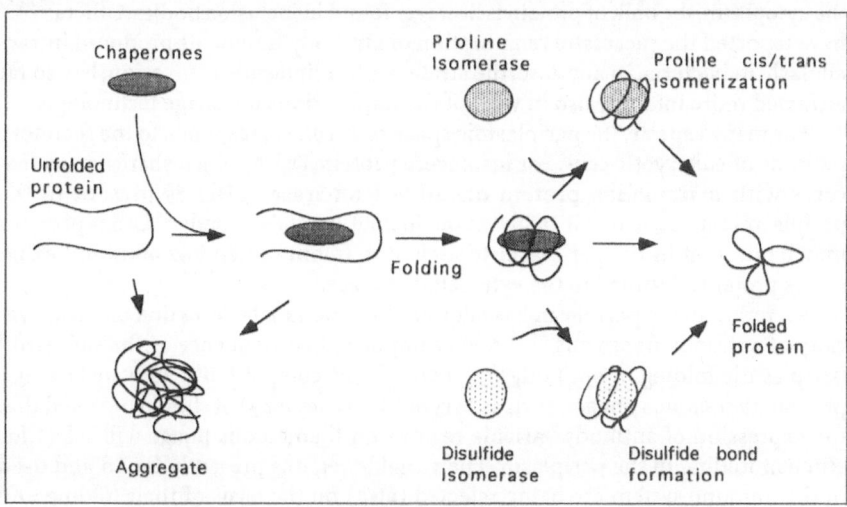

Fig. 4.4. Simplified overview of protein folding in *E. coli.*

An Artificial Environment:
Expressing Antibodies in Prokaryotes

The ultimate artificial environment is the test tube. It has been known for a long time that denatured antibodies can refold in vitro and regain their specific binding properties.[13,14,56,57] Nevertheless, antibodies in B cells and in plasma cells do receive help, by proteins such as BiP, GRP94 and PDI, to attain their functional tertiary and quaternary structure. For instance, in the endoplasmic reticulum, where the disulfide bond formation of antibodies takes place, this important reaction is catalyzed by protein disulfide isomerase (PDI), which interacts cotranslationally with nascent immunoglobulin chains. In order to achieve correct folding of its own proteins, *E. coli* contains a number of helper proteins, collectively referred to as "folding modulators", that catalyze certain folding steps (folding catalysts) or prevent the competing reaction of aggregation (see Fig. 4.4).

Antibodies or antibody fragments are increasingly being expressed in *E. coli* (and, as a subset, in *E. coli* bacteriophages). The main aim of this ectopic expression system is that of producing large amount of useful antibodies and of selecting new antibody specificities, exploiting the possibility of introducing rational or random modifications of many sorts including tags and fusion proteins of different nature. More relevant for this chapter, antibody expression in *E. coli* can provide important information on the process of antibody folding in artificial environments.

Three basic strategies can be used to express antibodies, and smaller fragments derived, in *E. coli:* (i) as inclusion bodies in the cytoplasm; (ii) as inclusion bodies in the periplasmic space or (iii) as soluble proteins secreted in the periplasm, under conditions in which periplasmic inclusion bodies are not formed and the cell wall is induced to leak.

At very high expression levels and low temperature, some functional Fab fragment can assemble in a soluble form in the cytoplasm of *E. coli.*[61] While this shows that antibody assembly can occur notwithstanding the reducing environment of the cytoplasm, the bulk of protein is however found in inclusion bodies. Others[58,63,64] have reported the successful renaturation of antibody fragments produced intracellularly in bacteria. In any case, the strategy of periplasmic expression, has so far attracted more interest, also in view of the implications for phage technology.

For many aspects, the periplasmic space of *E. coli* corresponds to the secretory pathway of eukaryotic cells. For instance, a protein, DsbA, which shares many features with mammalian protein disulfide isomerase (PDI), is present in the periplasmic space, where it catalyses the formation of S-S bonds.[64] Other proteins may be present in the periplasm to assist the folding of endogenous and exogenous proteins destined to the extracellular space.

Secretion in the periplasm has allowed the functional expression of a wide variety of antibody fragments.[65,66] A very important consequence of the successful periplasmic folding of antibody fragments is the compatibility with surface display on filamentous phages, such as M13 or fd [67] (chapter 3). It should be noted that the expression of antibody variable regions on filamentous phage will select for efficient folding in the periplasm. The variable regions presently found and used in the immune system are being selected (also) on the basis of their folding and quality control in B cells and plasma cells. By expressing V region repertoires in replication packages different from the B cell, one can probably select V regions suitable for folding in different cellular contexts. Philamentous phage are a first example of this, but it is conceivable that as the applications of ectopic antibody expression will increase, there will be the need to design schemes to select, from antibody repertoires, families of variable region frameworks more suitable for expression (and folding) in different cellular environments (for instance selection for variable regions that do not need the intrachain disulfide bond to properly fold (see chapter 10).

The folding of antibody fragments expressed in the periplasm does not proceed quantitatively, for most antibody fragments, and a high, albeit variable, proportion of it aggregates in an insoluble form.[68,69] The extent of aggregation depends in a crucial (yet still unpredictable) way on the primary sequence of the variable domains. Low bacterial growth temperature can reduce periplasmic aggregation, increasing the yield of folded antibody protein in the periplasm, but does not solve the problem.

The periplasmic folding process itself appears to limit the yield of functional protein in *E. coli*, leading to aggregation.[68,69] The following observations support this notion: (1) in the same vector, antibodies with different primary sequences differ dramatically in the distribution between periplasmic soluble and periplasmic precipitated protein; (2) variations in promoter strength usually increase the amount of periplasmic precipitated protein, but not of folded protein. Two slow steps of antibody folding in the periplasm of *E. coli* could conceivably be rate limiting: disulfide formation and proline cis-trans isomerization. Both steps occur during the folding process of antibody fragments and, for both, periplasmic proteins exist in *E. coli* that are known to catalyze these processes: proline cis-trans isomerase (rotamase) and disulfide isomerase (DsbA). Nevertheless, the overexpression of neither was found to have an effect on the folding of coexpressed antibody fragments, nor was there any synergistic effect noted when both chaperones were coordinately overexpressed.[69] Conversely, functional antibody fragments are not produced in the periplasm of DsbA deletion *E. coli* mutants and expression of plasmid encoded DsbA restores Fv formation.[69] Therefore, while periplasmic

DsbA is required for antibody fragment assembly, its functional over-expression does not significantly change the folding limit. Thus, proline cis-trans isomerase and disulfide isomerase appear not to be limiting under the conditions examined.

The role of different folding modulators was studied in vitro as well. Under in vitro refolding experiments, PDI and bacterial cytoplasmic chaperones (DnaK and GroEL/S) were shown to increase the yield of folded antibody fragments.[58] Also, in vitro refolding experiments[59] showed that the refolding of the L chain of a monoclonal antibody, which is limited by cis-trans isomerization, can be catalyzed with prolyl-cis-trans isomerase (PPIase) in vitro. In a very recent study[60] the effect of protein disulfide isomerase (PDI) and of bacterial folding modulators on ScFv antibody fragments synthesized in a cell-free translation system was investigated. This study demonstrated that the inclusion of PDI, but not of bacterial chaperones, leads to a 3-fold increase in the yield of functional antibody fragment over that obtained in the presence of glutathione redox system. The combination of PDI and chaperones increases the yield five-fold, while the disulfide-forming catalyst DsbA, alone, had no effect. Since eukaryotic PDI is known to catalyze disulfide isomerization (as opposed to net oxidation) much more efficiently than DsbA, this indicates that disulfide shuffling, rather than net formation, is the crucial yield-limiting step for antibody folding in these conditions. A direct chaperone-like effect of PDI on ScFv fragments, preventing their premature aggregation, cannot be excluded, but the fact that bacterial chaperones are ineffective argues against this. In any event, it is noteworthy that around 10% of the synthesized ScFv molecules was found to fold in an active state anyway, with no further additions. This may be very relevant for interpreting results obtained with the cytosolic expression of antibodies (see chapter 7).

In short, there is not yet a simple and general recipe for preventing protein aggregation in the periplasm, a phenomenon which occurs either before disulfide formation and proline cis-trans isomerization or at least independent of the extent of these reactions. This would be consistent with the idea that the formation of disulfide bonds occurs late in folding, when the proteins have already folded so as to place those cysteines that are to participate in disulfide bond formation in close proximity. Indeed, only a few cases are known in which assembly or polymerization depend on the actual formation of an intermolecular S-S bond. In most instances, the covalent bond does not cause dimerization or assembly, but merely stabilizes quaternary structures held together by noncovalent forces. Likewise, in general, the formation of intrachain disulfide bonds does not specify the native conformation, but simply stabilizes it.[70]

In any case, however, it appears that the major factor determining the folding efficiency of antibody fragments in *E. coli* is represented by the primary sequence of each antibody. Thus, the most effective strategy to improve folding and to obtain high-yield folding of periplasmic antibodies is to engineer the antibody itself, by changing its primary sequence. In a remarkable study, Knappik and Pluckthun[71] compared the framework sequences of one well expressed and one poorly expressed antibody, and were able to identify mutations, located in turns of the protein, which reduce the formation of aggregates during in vivo folding or which influence cell stability during expression. The two effects are based on different mutations and could be separated, but both mutations act synergistically in vivo. Neither mutation increases the thermodynamic stability in vitro. However, the in vivo folding mutation correlates with the yield of oxidative folding in vitro, which is limited by the side reaction of aggregation. This analysis shows that it is possible to engineer improved frameworks with increased folding properties and that limitations in recombinant antibodies expression can be overcome by single or multiple amino acid substitutions in framework residues.

Intrachain disulfide bonds are one of the hallmarks of the antibody domain architecture. In variable domains, they connect the two β sheets of the Ig domain, from strand b to strand f. The contributing cysteine residues L23 and L88 in VL and H22 and H92 in VH (numbering according to ref. 72) are almost perfectly conserved in all antibodies, as only 17 out of 5605 human and mouse VH sequences, and 18 out of 2985 VL sequences lack one of the cysteine residues. Since every known human or mouse germ line V region contains both cysteine residues, the observed cases are almost certainly due to somatic mutation. One of the reasons for this conservation appears to be the relatively low intrinsic stability of the V domain, which requires additional stabilization by the disulfide. The contribution of intrachain disulfide bonds to folding and stability of antibody variable domains was systematically studied for one antibody McPC603,[73] showing that both disulfide bonds (one in VH and one in VL) are necessary for the folding and stability of the Fv, of the Fab and of the single-chain Fv fragments. It may be surmised that domains which are intrinsically more stable may tolerate the removal or the absence of the disulfide and, conversely, that those that do tolerate removal may be intrinsically more stable than average.

The first antibody with a naturally missing disulfide that has been investigated for folding and stability[74] is the leaven binding antibody ABPC48, in which the second half-cysteine in the VH is substituted by a tyrosine residue.[75] Hence, the presence of a disulfide bond in VH does not appear to be necessary for the function of this antibody and may not be required for folding of the Ig domain in general, as has been assumed so far on the basis of the almost absolute evolutionary conservation of the two cysteine residues in the entire Ig superfamily. The ScFv version of the ABPC48 antibody was expressed in *E. coli*, together with a mutant in which the VH disulfide bond was restored, and the stability of the ABPC48 and disulfide-restored mutants was compared.[74] While the ABPC48 protein was found to be less stable than an average ScFv molecule, the restored disulfide increased its stability well above that of other ScFv fragments, explaining why it tolerates the disulfide loss. Surprisingly, the unpaired cysteine residue appears to be solvent exposed, in contrast to the deeply buried disulfide bond of ordinary variable domains. This implies a very unusual conformation of strand b containing the unpaired cysteine Cys H22, which might be stabilized by interactions with the tyrosine residue in position H92, thus compensating for the loss of free energy from the missing disulfide.

In another study,[76] two amino acid mutations (Y32H and C23V) were introduced sequentially into the human VK domain REI$_v$. The first change stabilizes the folded state of the domain by 4.6 kJ/mol, while the second one, which abolishes the central disulfide bridge, destabilizes the folded domain by 17.5 kJ/mol. Introduction of the stabilizing mutation first is a necessary prerequisite to the removal of the central disulfide bridge without collapse of the fold. The double mutant VK domain can be accumulated in a functional form in the cytoplasm of *E. coli*.

Thus, it appears that variable domains have a range of folding stabilities and that the overall stability of the fold is contributed by many critical residues or combinations of residues in the framework regions. The loss, or the absence, of the cysteine residues will be therefore tolerated or not, according to the overall stability of each individual variable domain.

The cell cytoplasm is reducing.[77] The cytoplasmic expression of two different ScFv fragments was studied[78] in strains of *E. coli* in which the thioredoxin reductase trxB gene was deleted. In these mutants, the formation of disulfide bonds in the cytoplasm was found to be enhanced.[79] Interestingly, much higher amounts of functional, disulfide-containing ScFv could be produced in the cytoplasm of *E. coli* in the trxB- strain than in the trxB+ strain.[78] The trxB null mutation has no or only very little influence on the total amount of soluble protein, but does dramatically increase the amount of disulfide-containing antibody in the cytoplasm. Thus, such systems might be especially attractive for screening and selection systems for cytosolic antibodies.

Notwithstanding the successful examples of rational or semirational improvement of the folding stability of antibodies, it is difficult to identify a priori, in a general way, those amino acid replacements that are likely to induce stabilization of the protein, even in cases of known three-dimensional structure. Short of a guiding rationale based on first principles of protein structural chemistry, an alternative possibility is to use molecular repertoire technology and a genetic screen for folding stability. A system along these lines was recently described[80] exploiting the ToxR-mediated signal transduction pathway of *Vibrio cholera*,[81] which links protein dimerization in the periplasmic space to transcriptional activation. Using this bacterial signal transduction system, a model repertoire of variants of the immunoglobulin VK domain REI$_V$, fused to ToxR, was genetically screened for folding stability, and it was shown that differences in transcription of a reporter gene correlate positively with folding stabilities of the corresponding REI$_V$ mutants.

Conclusion

In conclusion, the proper assembly of antibodies in the secretory pathway of lymphoid, and to a certain extent, nonlymphoid mammaliancells, is a highly regulated process that undergoes an extensive process of quality control. Proteins interacting with the constant regions play a major role in this process. The expression of smaller antibody domains, such as ScFv fragments, highlights the influence of residues within the variable regions on the performance of antibody fragments ectopically expressed. Studies in *E. coli* start to show how rational design of "super-frameworks", cell engineering or selection schemes can be exploited to improve the folding of ectopically expressed antibodies.

Acknowledgments

We are very grateful to Roberto Sitia for his contribution to the first part of this chapter, which is based extensively on reference 82.

References

1. Anfinsen CB. Principles that govern the folding of protein chains. Science 1973; 181:223-230.
2. Gething MJ, Sambrook J. Protein folding in the cell. Nature 1992; 355:33-45.

3. Hurtley SM, Helenius A. Protein oligomerization in the endoplasmic reticulum. Ann Rev Cell Biol 1989; 5:277-307.
4. Pelham HBR. Control of protein exit from the endoplasmic reticulum. Annu Rev Cell Biol 1989; 5:1-23.
5. Klausner RD, Sitia R. Protein degradation in the endoplasmic reticulum. Cell 1990; 62:611-614.
6. Fra AM, Fagioli C, Finazzi D et al. Quality control of ER synthesized proteins: an exposed thiol group as a three-way switch mediating assembly, retention and degradation. EMBO J 1993; 12:4755-4761.
7. Bonifacino JS, Lippincott-Schwartz J. Degradation of proteins within the endoplasmic reticulum. Curr Opin Cell Biol 1991; 3:592-600.
8. Fra AM, Sitia R. The endoplasmic reticulum as a site of protein degradation. In: Borgese N, Harris JR eds. Subcellular Biochemistry. Vol 21. New York: Plenum Press, 1993:143-168.
9. Bergman LW, Kuehl WM. Formation of intermolecular disulfide bonds on nascent immunoglobulin polypeptydes. J Biol Chem 1979; 254:5690-5694.
10. Bergman LW, Kuehl WM. Formation of an intrachain disulfide bond on nascent immunoglobulin light chain. J Biol Chem 1979; 254:8869-8876.
11. Ooi CE, Weiss J. Bidirectional movement of a nascent polypeptide across microsomal membranes reveal requirements for vectorial translocation of proteins. Cell 1992; 71:87-96.
12. Haber E. Recovery of antigenic specificity after denaturation and complete reduction of disulfides in a papain fragment of antibody. Proc Natl Acad Sci USA 1964; 52:1099-1106.
13. Whitney P, Tankford C. Recovery of specific activity after complete unfolding and reduction of an antibody fragment. Proc Natl Acad Sci USA 1965; 53:524-532.
14. Valetti C, Sitia R. The differential effects of dithiothreitol and 2-mercaptoethanol on the secretion of partially and completely assembled immunoglobulins suggest that thiol-mediated retention does not take place in or beyond the Golgi. Mol Bio Cell 1994; 5:1311-1324.
15. Seligmann ME, Preudhomme JL et al. Heavy chain disease: current findings and concepts. Immunol Rev 1979; 48:145-167.
16. Haas IG, Wabl M. Immunoglobulin heavy chain binding protein. Nature 1983; 306:387-389.
17. Bole DG, Hendershot LM et al. Post-traslational association of immmunoglobulin heavy chain binding protein with nascent heavy chains in nonsecreting and secreting hybridomas. J Cell Biol 1986; 102:1558-1566.
18. Haas IG. BiP-A heat shock protein involved in immunoglobulin chain asssembly. Curr Top Microb Immunol 1991; 167:71-82.
19. Flynn GC, Pohl J, Flocco MT et al. Peptide-binding specificity of the molecular chaperone BiP. Nature 1993; 353:726-730.
20. Melnick J, Dul JL, Argon Y. Sequential interacion of the chaperones BiP and GRP94 with immunoglobulin chains in the endoplasmic reticulum. Nature 1994; 370:373-375.
21. Frydman J, Nimmesgern E, Ohtsuka K et al. Folding of nascent polypeptide chains in a high molecular mass assembly with molecular chaperones. Nature 1994; 370:111-117.
22. Munro S, Pelham HRB. A C-terminal signal prevents secretion of luminal ER proteins. Cell 1987; 48:899-907.
23. Hendershot L, Bole D, Kohler G et al. Assembly and secretion of heavy chains that do not associate postranslationally with immunoglobulin heavy chain-binding protein. J Cell Biol 1987; 104:761-767.

24. Hendershot LM. Immunoglobulin heavy chain and binding protein complexes are dissociated in vivo by light chain addition. J Cell Biol 1990; 111:829-837.
25. Hamers-Casterman C, Atarhouch T, Muyldermans S et al. Naturally occurring antibodies devoid of light chains. Nature 1993; 363:446-448.
26. Secher DS, Milstein C, Adetugbo K. Somatic mutants in antibody diversity. Immunol Rev 1977; 36:51-72.
27. Dul JL, Argon Y. A single amino acid substitution in the variable region of the light chain specifically block immunoglobulin secretion. Proc Natl Acad Sci USA 1990; 87:8135-8139.
28. Gardner AM, Aviel S, Argon Y. Rapid degradation of an unassembled immuno-globulin light chain is mediated by a serine protease and occurs in a pre-Golgi compartment.J Biol Chem 1993; 268:25904-25947.
29. Knittler MR, Haas IG. Interaction of BiP with newly synthesized immunoglobulin light chain molecules: cycles of sequential binding and release. EMBO J 1992; 11:1573-1581.
30. Weiss A, Littman DR. Signal transduction by lymphocyte antigen receptors. Cell 1994; 76:263-274.
31. Sitia R, Neuberger MS, Milstein C. Regulation of membrane IgM expression in secretory B cells: translational and post-translational events. EMBO J 1987; 6:3969-3977.
32. Hombach J, Sablitzky F, Rajwsky K et al. Transfected plasmacytoma cells do not transport the membrane form of IgM to the cell surface. J Exp Med 1988; 167:652-657.
33. Moller G. The B cell antigen receptor. In: Moller G ed. Immunol Rev Vol 132. Copenhagen: Munksgaard, 1993.
34. Sitia R, Rubertelli A, Kikutani H et al. The regulation of membrane bound and secreted alpha chain biosynthesis during differentiation of the I.29 B cell lymphoma. J Immunol 1985; 135:289-2864.
35. Sitia R, Alberini C, Biassoni R et al. The control of membrane and secreted heavy chain biosynthesis varies in different immunoglobulin isotypes produced by a monoclonal B cell lymphoma. Mol Immunol 1988; 25:189-197.
36. Sitia R, Neuberger MS, Alberini CM et al. Developmental regulation of IgM se-cretion: the role of the carboxy-terminal cysteine. Cell 1990; 60:781-790.
37. Davis AC, Roux KH, Pursey J et al. Intermolecular disulfide bonding in IgM: ef-fects of replacing cysteine residues in the μ heavy chain. EMBO J 1989; 8:2519-2526.
38. Randall TD, Brewer JW, Corley RB. Direct evidence that J chain regulates the polymeric structure of IgM in antibody-secreting B cells. J Biol Chem 1992; 267:18002-18007.
39. Randall TD, Parkhouse RME, Corley RB. J chain synthesis and secretion of hexameric IgM is differentially regulated by lipopolysaccharide and interleukin 5. Proc Natl Acad Sci USA 1992; 89:962-966.
40. Raschke WC, Mather EL, Koshland ME. Assembly and secretion of pentameric IgM in a fusion beetween a nonsecreting B cell lymphoma and an IgG-secreting plasmacytoma. Proc Natl Acad Sci USA 1979; 76:3469-3473.
41. Mather EL, Alt FW, Bothwell ALM et al. Expression of J chain RNA in cell lines representing different stages of B lymphocyte differentiation. Cell 1981; 23:369-378.
42. Cattaneo A, Neuberger MS. Polymeric IgM is secreted by transfectants of nonlymphoid cells in the absence of immunoglobulin J chain. EMBO J 1987; 6:2753-2758.
43. Alberini CM, Bet P, Milstein C et al. Secretion of immunoglobulin M assembly intermediates in the presence of reducing agents. Nature 1990; 347:485-487.

44. Guenzi S, Fra AM, Sparvoli A et al. The efficiency of cysteine-mediated intracellular retention determines the differential fate of secretory IgA and IgM in B and plasma cells. Eur J Immunol 1994; 24:2477-2482.

45. Reddy P, Sparvoli A, Fagioli C et al. Formation of reversible disulfide bonds with the protein matrix of the endoplasmic reticulum correlates with the retention of unassembled Ig light chains. EMBO J 1996; 15:2077-2085.

46. Piccioli P, Ruberti F, Biocca S et al. Neuroantibodies: molecular cloning of a monoclonal antibody against substance P for expression in the central nervous system. Proc Natl Acad Sci USA 1991; 88:5611-5615.

47. Piccioli P, Di Luzio A, Amann R et al. Neuroantibodies: ectopic expression of a recombinant anti-substance P antibody in the central nervous system of transgenic mice. Neuron 1995; 15:1-12.

48. Düring K, Hippe S, Kreuzaler F et al. Synthesis of a functional monoclonal antibody in transgenic *Nicotiana tabacum*. Plant Mol Biol 1990; 15:281-293.

49. Williams GT, Venkitaraman AR, Gilmore DJ et al. The sequence of the μ transmembrane segment determines the tissue specificity of the transport of immunoglobulin M to the cell surface. J Exp Med 1990; 17:947-952.

50. Biocca S, Ruberti F, Tafani M et al. Redox state of single chain Fv fragments targeted to the endoplasmic reticulum, cytosol and mitochondria. Bio/Technology 1995; 13:1110-1115.

51. Biocca S, Cattaneo A. Intracellular immunization: antibody targeting to subcellular compartments. Trends in Cell Biology 1995; 5:248-252.

52. Biocca S, Tafani M, Cattaneo A Assembled IgG molecules are exported from the endoplasmic reticulum in myeloma cells despite the retention signal SEKDEL. Submitted.

53. Carayannopoulos L, Max EE, Capra JD. Recombinant IgA expressed in insect cells. Proc Natl Acad Sci USA 1994; 91:8348-8352.

54. Ma JK-C, Lehner T, Stabila P et al. Assembly of monoclonal antibodies with IgG1 and IgA heavy chain domains in transgenic tobacco plants. Eur J Immunol 1994; 24:131-138.

55. Ma JK-C, Hiatt A, Hein M et al. Generation and assembly of secretory antibodies in plants. Science 1995; 268:716-719.

56. Sears DW, Mohrer J, Beychok S. A kinetic study in vitro of the reoxidation of interchain disulfide bonds in a human immunoglobulin IgG1k. Correlation between sulphydryl disappearance and intermediates in covalent assemblyof H2L2. Proc Natl Acad Sci USA 1975; 72:353-357.

57. Kabat EA. Structural concepts in immunology and immunochemistry. In: Holt, Rinehart and Wiston eds. Molecular and cellular biology series. New York, 1968.

58. Buchner J, Brinkmann U, Pastan I. Renaturation of a single-chain immunotoxin facilitated by chaperones and protein disulfide isomerase. Bio/Technology 1992; 10:682-685.

59. Lang K, Schmid FX, Fischer G. Catalysis of protein folding by prolyl isomerase. Nature 1987; 329:268-270.

60. Ryabova LA, Desplancq D, Spirin AS et al. Functional antibody production using cell-free translation: effects of protein disulfide isomerase and chaperones. Nature Biotechnology 1997; 15:79-84.

61. Cabilly S. Growth at suboptimal temperatures allows the production of functional, antigen binding Fab fragments in *Escherichia Coli*. Gene 1989; 85:553-557.

62. Carter P, Kelley RF, Rodrigues ML et al. High level of *Escherichia Coli* expression and production of a bivalent humanized antibody fragment. Biotechnology 1992; 10:163-167.

63. Duenas M, Vazquez, Ayala M et al. Intra- and extracellular expression of an ScFv antibody fragment in *E. coli*: effect of bacterial strains and pathway engineering using GroES/L chaperonins. BioTechniques 1994; 16:476-483.
64. Bardwell JCA, Beckwith J. The bonds that tie: catalyzed disulfide bond formation. Cell 1993; 74:769-771.
65. Skerra A, Plückthun A. Assembly of a functional immunoglobulin Fv fragment in *E. coli*. Science 1988; 240:1038-1041.
66. Plückthun A. Mono- and bivalent antibody fragments in *Escherichia coli*: engineering, folding and antigen binding. Immunol Rev 1992; 130:151-188.
67. Winter G, Griffiths AD, Hawkins RE et al. Making antibodies by phage display technology. Ann Rev Immunol 1994; 12:433-455.
68. Skerra A, Plückthun A. Secretion and in vivo folding of the Fab fragment of the antibody McPC603 in *Escherichia coli*: influence of disulfides and cis-prolines. Protein Engineering 1991; 4:971-979.
69. Knappik A, Krebber C, Plückthun A. The effect of folding catalysts on the in vivo folding process of different antibody fragments expressed in *Escherichia coli*. Bio/ Technology 1993; 11:77-83.
70. Creighton TE. Protein folding. Biochem J 1990; 270:1-16.
71. Knappik A, Plückthun A. Engineered turns of a recombinant antibody improve its in vivo folding. Protein Engineering 1995; 8:81-89.
72. Kabat EA, Wu TT, Perry HM et al. Variable region hevy chain sequences. In: National Technical Information Service (NTIS). Sequences of Proteins of Immunological Interest. NIH Publication No. 91-3242. 1991.
73. Glockshuber R, Schmidt T, Plückthun A. The disulfide bonds in antibody variable domains:effects on stability, folding in vitro and functional expression in *Escherichia cli*. Biochemistry 1992; 31:1270-1279.
74. Proba K, Honegger A, Plückthun A. A natural antibody missing a cysteine in VH: consequences for thermodynamic stability and folding. J Mol Biol 1997; 265:161-172.
75. Rudikoff S, Pumphrey JG. Functional antibody lacking a variable region disulfide bridge. Proc Natl Acad Sci USA 1986; 83:7875-7878.
76. Frisch C, Kolmar H, Fritz HJ. A soluble immunoglobulin variable domain without a disulfide bridge: construction, accumulation in the cytoplasm of *E. coli*, purification and physicochemical characterization. Biol Chem Hoppe Seyler 1994; 375:353-356.
77. Hwang C, Sinskey AJ, Harvey FL. Oxidized redox state of glutathione in the endoplasmic reticulum. Science 1992; 257:1496-1502.
78. Proba K, Ge L, Plückthun A. Functional antibody single-chain fragments from the cytoplasm of *Escherichia coli*: influence of thioredoxin reductase (TrxB). Gene 1995; 159:203-207.
79. Derman AI, Prinz WA, Belin D et al. Mutations that allow disulfide bond formation in the cytoplasm of *Escherichia Coli*. Science 1993; 262:1744-1747.
80. Kolmar H, Frisch C, Goetze K et al. Immunoglobulin mutant library genetically screened for folding stability exploiting bacterial signal transduction. J Mol Biol 1995; 251:471-476.
81. DiRita VJ. Coordinate expression of virulence genes by ToxR in Vibrio cholerae. Mol Microbiol 1992; 6:451-458.
82. Sitia R, Catrones A. Synthesis and assembly of antibodies in natural and artificial environments. In: The antibody I, Zonetti N and Capra JD Ed. 1995:127-168.

CHAPTER 5

Protein Sequence Motifs Involved in Intracellular Trafficking

Silvia Biocca and Antonino Cattaneo

Major advances have occurred in our understanding of the intracellular targeting of proteins in eukaryotic cells. Many proteins harbor short and conserved amino acid motifs, often separate from other functional parts of the molecule that are involved in their intracellular targeting. These sequences, known as targeting signals, can determine the spatial location of a protein, its folding, post translational modification and degradation. The conservation of these motifs across evolution varies: in some cases they are highly conserved, while in others, substitutions of few amino acids are tolerated, provided there is no change in a certain pattern of charge or hydrophobicity of the motif. Most often, these targeting motifs are autonomous and dominant, in that, if grafted onto a reporter protein, they are able to independently specify its targeting.

The existence of these linear motifs represents the basis of the strategy of intracellular antibodies. The possibility of targeting antibodies and/or antibody forms to different intracellular compartments by incorporating specific localization signals, is a task strictly connected with the evolving field of the cell biology of intracellular targeting, compartmentalization and degradation of proteins.

This chapter will seek to describe some examples of transplantable targeting signals which have been shown to be necessary to the newly synthesized proteins for sorting out their correct location within the secretory pathway for entering the cytosol, nucleus, nucleolus or mitochondria or for being targeted for destruction. This is not a review on the molecular mechanisms of intracellular protein targeting, but rather a description of those targeting signals that have or could be usefully and directly applied to target antibodies and antibody domains to different intracellular compartments.

Sorting in the Secretory Pathway

Historically, the first localization sequence to be characterized was the hydrophobic signal sequence for secretion, which is found at the N-terminus of proteins that enter the endoplasmic reticulum (ER).[1,2] During the biosynthesis of these proteins, the signal sequence allows the corresponding mRNA to be recognized by polyribosomes bound to intracellular membranes and is cleaved by specific peptidases in the ER lumen after the nascent polypeptide has been cotranslationally inserted through the ER membrane.[3] The presence or the absence of the leader

Intracellular Antibodies: Development and Applications, edited by Antonino Cattaneo and Silvia Biocca. © Springer-Verlag and Landes Bioscience 1997.

sequence will determine the first "decision" on the location of a protein: secretory pathway versus cytosol, that is, translation by membrane bound versus free ribosomes.

The secretory pathway of eukaryotic cells can be considered to start in the ER. The newly synthesized proteins sequentially pass through a series of different membrane-bound compartments that show distinct protein compositions. Some of these compartments are schematically depicted in Figure 5.1. Morphologically, it is difficult to define the boundary between the ER and the Golgi complex due to the dynamic nature of the two compartments.[4] Some controversy still exists in the definition of the distinct functional subcompartments of the Golgi complex (for a review see refs. 5-8), but one can define the Golgi intermediate compartment (ERGIC), the cis Golgi network (CGN), the cisternae of the Golgi stack (GS) and the trans Golgi network (TGN). Within the TGN, newly synthesized proteins undergo terminal modifications in the polypeptide and carbohydrate chains, such as proteolytic cleavage, addition of sialic acid and/or sulfate groups. The TGN is a crossing point, not only for proteins directed to cellular compartments such as plasma membrane, lysosomes, endosomes and regulated secretory granules, but also for proteins recycled from the plasma membrane and endosomal compartments after their internalization by the process of endocytosis.[5]

The sorting activity of the exocytic pathway, that is the mechanism by which all the membrane compartments acquire the correct pattern of proteins, is mostly achieved through the sequential budding, targeting and fusion of small (50-80 nm) vesicles. Important progress in defining the components involved in vesicles budding and fusion has been made recently, mostly by Rothmann and colleagues.[6,7] A series of polypeptides necessary for budding has been identified as coat proteins (COPs) and the ADP-ribosylation factor (ARF). Once a transport vesicle has formed, it must find the correct acceptor membrane with which to fuse. The specificity of targeting and fusion is instrumental to the vectorial and precise nature of intracellular transport. Small GTP-binding proteins, called Rab proteins, and a class of cytoplasmically oriented membrane proteins, called SNAREs, play a key role in targeting and fusion of vesicles. At each stage of vesicular transport, both vesicles and the target membrane would have their own unique SNAREs: v-SNAREs, localized in the transport vesicles and t-SNARE, present in the target membranes. Their interaction would mediate the process of fusion between vesicles. Thus, the correct localization of a soluble or a membrane protein along the secretory pathway is achieved by a superposition of two targeting mechanisms: the targeting of proteins into selected classes of vesicles and the targeting of the vesicles themselves.

The mechanisms and components involved in vesicular transport has been extensively reviewed (see ref. 8 and references therein) and will not be dealt with further.

Vesicle-mediated secretion from animal cells utilizes three types of secretory vesicles. The first type is the constitutive secretory vesicle (CSV), which mediates the constitutive continuous secretion of newly synthesized proteins. In polarized cells, such as epithelial and neurones, two different types of CSV coexist: CSVs which deliver proteins to the apical/axonal and basolateral/dendritic domains of the plasma membrane. The second type is the secretory granule in neurones, also referred to as the dense core vesicle, which occurs only in cells capable of regulated protein or peptide secretion and mediates the stimulus-dependent release of stored

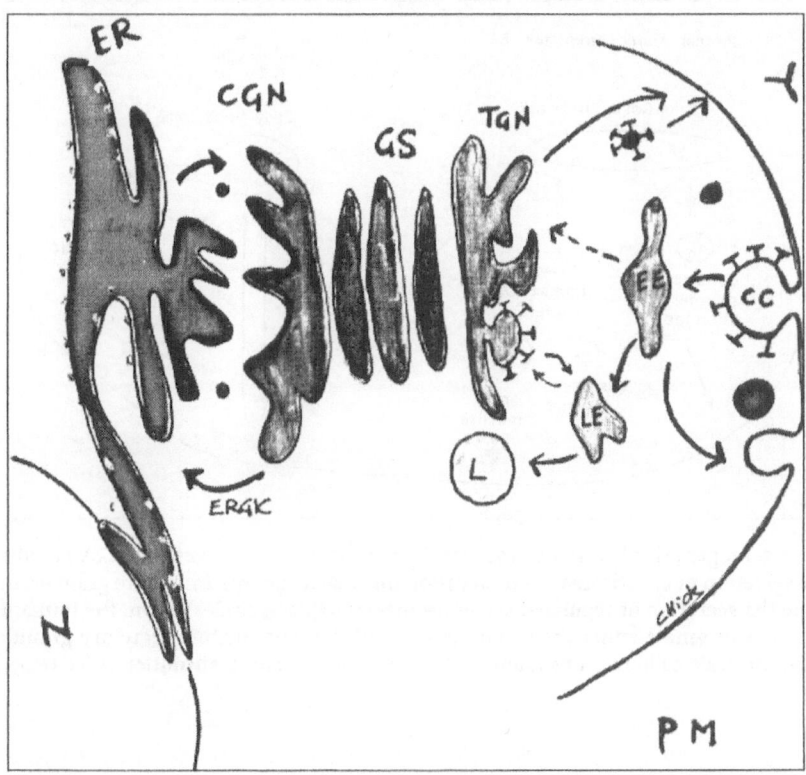

Fig. 5.1. Membrane systems in the secretory pathway. The membrane system depicted in the scheme comprises: the ER, the nuclear membrane (N), the ER to the Golgi intermediate compartment (ERGIC), the cis Golgi network (CGN), the cisternae of the Golgi stack (GS), the trans Golgi network (TGN), the early endosomes (EE), the late endosomes (LE), the lysosomes (L), the clathrin-coated vesicles (CC) and the plasma membrane (PM). The arrows indicate the traffic and membrane recycling pathways.

proteins. The third type is the synaptic vesicle of neurones (and their endocrine counterpart, the synaptic-like microvesicles), which mediates the release of neurotransmitters, but lacks proteins. While the first two classes of vesicles derive from the TGN, the biogenesis of synaptic vesicles occurs from the early endosome[9] (Fig. 5.2).

A distinct set of shuttle vesicles would mediate the transport of membrane proteins along the secretory or endocytic pathway. Thus, the intracellular location of membrane proteins follows the intracellular trafficking of a corresponding set of vesicles. For a given protein, entering or not entering a selected shuttle vesicle is determined by the presence or absence of multiple sorting signals, each specifying the fate of that protein at successive stages of maturation. The combination of different sorting signals will determine the itinerary of the protein inside the cell.

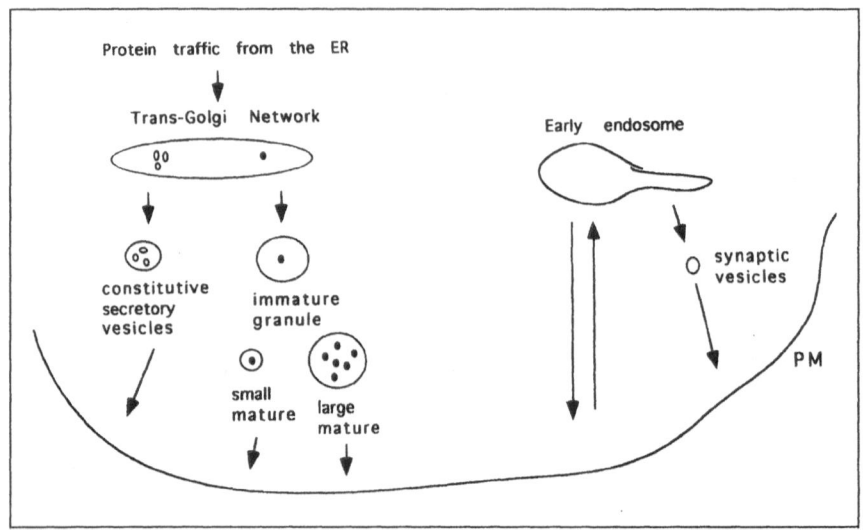

Fig. 5.2. Biogenesis of secretory vesicles. Constitutive secretory vesicles (CSV) mediate the secretion of constitutive secretory proteins (black square). Immature granules mediate the secretion of regulated secretory proteins (black circles). From the immature granules originate either the small mature and the large mature secretory granules. Synaptic vesicles in neurons mediate the secretion of neurotransmitters. (See ref. 9.)

One can envisage a protein being actively concentrated in a given vesicle by a sorting signal, or, vice versa, a protein being restricted to enter a given vesicle by a specific retention signal.

According to the bulk-flow model,[10-12] export from the ER occurs mainly through an indiscriminate pathway, in which proteins lacking specific export signals diffuse into budding vesicles to be transported to the next compartment. However, a large body of evidence questions this theory and suggests the existence of tissue- or cell-specific mechanisms also for the constitutive secretion.[13-15]

Most proteins are matured as they progress through the compartments of the secretory pathway, interacting with resident ancillary proteins and enzymes involved in a complex quality control system (see chapter 4).[16] Unassembled or misfolded subunits of oligomeric proteins are recognized and targeted towards an ER degradative compartment.[17,18] These properties require a fine discrimination between resident proteins, those to be exported and those targeted for ER degradation. This is achieved by the simultaneous presence of multiple localization signals on a given oligomeric protein. This is true, for example, for the T cell antigen receptor, where a short stretch of amino acids in the transmembrane domain of the α subunit functions as a retention signal and as a determinant for the rapid degradation only when this subunit is not assembled in a functional T cell receptor complex.[19]

The coexistence of different signals in a single oligomeric protein has been recently observed, for example, for the immunoglobulin E Fc receptor. A combination of more signals present in different subunits, which are subsequently exposed or masked during biogenesis in the ER, guarantees the correct maturation of the protein.[20]

The mechanisms of degradation present in the ER are only partially known, but are distinct from the degradation system based on lysosomes. Recently, a direct retrotranslocation of proteins targeted for degradation from the ER to the cytosol has been observed. Once in the cytosol, these proteins are poly-ubiquitinated and degraded by proteasomes.[21-23]

The mannose-6-phosphate receptor binds glycoproteins bearing appropriately modified mannose residues on their N-linked glycans and delivers these proteins to the lysosomal compartments from the trans-Golgi network.[24] This is but one of a growing number of examples in which the presence of a specific sugar moiety is recognized by a targeting receptor (tasting the sugar). Thus, glycosylation can function as a positive "exit signal" for membrane glycoproteins that reach the membrane by bulk flow, without specific retention signals.[25,26] One example to support such a function of oligosaccharides has been obtained using a recombinant chimeric membrane protein based on rat growth hormone. A membrane-anchored form was engineered and remained intracellular. The introduction of N-glycosylation sites resulted in its glycosylation and presentation on the cell surface.[27]

The detailed mechanisms regulating the intracellular trafficking in the exocytic pathway are still to be fully elucidated. Some of the sorting signals, however, have been characterized and when transplanted onto reporter proteins, are necessary and sufficient to redirect their location (Table 5.1).

Retention and Retrieval in Endoplasmic Reticulum and Golgi Apparatus

The KDEL Retention Signal

Resident integral membrane proteins and resident soluble proteins of the ER and the Golgi are subjected to retrograde transport, that allows them to escape exocytic transport vesicles. A carboxy-terminal peptide KDEL is present at the COOH-terminus of many resident lumenal proteins of the ER,[28-30] and constitutes a transport retrieval signal from the Golgi to the ER. These soluble proteins are, thus, continuously retrieved in the lumen of the ER from a post-ER "salvage" compartment.[31] Transplantation of the KDEL signal onto different reporter proteins prevents their secretion, resulting in their retention in the lumen of the ER.

Although the KDEL retention signal appears to be universal, this signal fails to efficiently retain some reporter proteins in a cell-type specific manner[32,33] (see chapter 4).

Other, more specific retention signals have been described, for example, the C-terminal cysteine residue of the μ chain of IgM, has been shown to play a critical role in intracellular retention of unpolymerized chains[34] (see chapter 4). Thiol-mediated retention mechanism, similar to those regulating IgM secretion, have been described for acetylcholinesterase, an enzyme secreted only as disulfide-linked homodimers[35] (see chapter 4).

Table 5.1. Examples of transplantable signals in the secretory pathway

Signal	Location in Protein	Specified Localization
KDEL	COOH-terminus, luminal	Retrieval of proteins from Golgi to ER
KKXX	COOH-terminus, in cytoplasm	Retrieval of membrane proteins from Golgi to ER
XXRR	N-terminus, in cytoplasm	Retrieval of membrane proteins from Golgi to ER
Transmembrane domain TMD	Transmembrane	Retention of membrane proteins in Golgi
Propeptide	N-terminus, luminal	Transport from Golgi to endosomes
Mannose 6-phosphate	Asn-linked, saccharides luminal	Transport from Golgi to endosomes
Tyrosine-rich dileucine	Cytoplasmic tail	Transport from Golgi to endosomes
YQRL	Cytoplasmic tail	Transport from cell surface to Golgi
YXXZ/NXXY	Cytoplasmic tail	Transport from cell surface to endosomes
YXXI	Cytoplasmic tail	Transport from Golgi to lysosomes
GPI-anchor	COOH-terminus, luminal	Transport from Golgi to apical cell surface in polarized cells

The Double Lysine Motif

Several integral proteins that are resident in the ER and in the ER-Golgi inter-mediate compartment, possess retrieval signals in their cytoplasmic tail. In resi-dent proteins with a type I topology (amino terminus in the lumen), the signal consists of two lysines at position -3 and -4 (KKXX) from the COOH-terminal end of the cytoplasmic domain.[36] In type II proteins (carboxyl terminus in the lumen), the signal consists of two critical arginines (RR) within the first five N-terminal amino acids of the proteins.[37]

The Transmembrane Domain Signal

In contrast to the above mentioned ER resident integral proteins, several Golgi-stack integral membrane proteins are retained through a retention signal located in the transmembrane domain (TMD)(for a review see refs. 38 and 39). No con-sensus sequence for retention has been found in these TMD, but it appears that the length in amino acids of this part of the molecule plays a crucial role in the reten-tion of the protein. Analysis of the effects of mutations in the TMD on retention of sialyltransferase (ST), an enzyme of the mammalian trans-Golgi, are more consistent with the TMD exerting its effect because of its length rather then for its ability to form protein-protein interactions. The addition or removal of residues showed that the efficiency of retention is related to TMD lenght.[40]

The transmembrane domain of the membrane form of IgM (μm) acts as an ER membrane retention signal in nonlymphoid cells[41] (see chapter 4).

Tyrosine-Containing Motifs

A tyrosine residue (in the tetrapeptide Tyr-Gln-Arg-Leu) in the cytoplasmic tail is critical to direct the integral membrane protein TGN38 to the TGN.[42,43] The tyrosine containing motif (for a review see ref. 44) is present in a variety of proteins destinate to the TGN and, in a 'correct context', functions as a cytoplasmic domain signal for the rapid internalization of integral membrane proteins through a mechanism based on clathrin-coated pits.[4,45] Deletion or point mutation of the tyrosine residue leads to the accumulation of TGN38 on the cell surface, suggesting that it normally works as a retrieval signal. The presence of a serine residue enhances appearance of TGN38 on the plasma membrane and in late endosomes, suggesting a mechanism of phosphorylation/dephosphorylation of this residue which may modulate recycling of proteins. This modulation of TGN retention by phosphorylation/dephosphorylation has been demonstrated for another TGN membrane protein, the furin processing enzyme.[46,47]

This tyrosine signal is one example of a large class of tyrosine based signals that mediate transmembrane protein sorting to different compartments in the endocytic and late secretory pathways (see Table 5.2), including lysosomes.

Targeting to Lysosomes

Targeting of membrane proteins to lysosomes is a very complex pathway. Newly synthesized proteins in the ER pass through the Golgi complex and upon arrival to the TGN, are directly targeted to endosomes or first, to the plasma membrane where they are rapidly internalized into the endocytotic pathway to the lysosomes. These proteins utilize the tyrosine[48] and leucine-based sorting motifs (YXXZ/NXXY and LZ where Z indicate one of the hydrophobic amino acids) in their cytoplasmic tails to select the TGN or the plasma membrane as the points of entry into the endosomal corridor to lysosomes.[49] The tyrosine-based and the dileucine motif are individually sufficient to induce both endocytosis and delivery to lysosomes, but chimeras containing both, were predominantly delivered to lysosomes.[50] The similarity of the sorting signals for the TGN resident proteins and these signals necessary for lysosome targeting is high and has been suggested that the correct spacing of these tetrapeptides relative to the membrane may serve as determinants for the differential sorting between the two compartments. This suggestion comes from the analysis of mutants of Lamp 1, a type I transmembrane glycoprotein that is localized primarily in lysosomes: the tyrosine-based signal YXXI of the cytoplasmic tail is necessary for the protein to be targeted to lysosomes and small changes in the spacing relative to the lipid bilayer almost completely blocks the trafficking to lysosomes.[51]

GPI-Anchor

Many eukaryotic cell-surface proteins are anchored to the cell membrane by carboxy-terminal linkage to glycosylphosphatidylinositol (GPI).[52,53] The GPI modification of proteins occurs post-translationally in the lumen of the ER. The GPI

Table 5.2. Examples of tyrosine-based sorting signals

Signal Sequence	Protein	Localization
NPXY-type		
TM-12aa-FDNPVY-32aa	LDL receptor	Early endosomes
TM-16aaSVNPEY-387aa	IGF-1 receptor	Early endosomes
TM-12aa-FENTLY-27aa	Mannose receptor	Phagosomes
YXXZ-type		
17-aaLSYTRF-45aa-TM	Transferrin receptor	Early endosomes
MTKEYQDL-32aa-TM	Asialoglycoprotein receptor H1	Early endosomes
TM-24aa-YKYSKV-135aa	CI Man-6-P receptor	Late endosomes
TM-42aa-AAYRGV-19aa	CD Man-6-P receptor	Late endosomes
TM-RKRSHAGYQTI	Lamp-1	Lysosomes
TM-KHHHAGYEQF	Lamp-2	Lysosomes
TM-RMQAQPPGYRHV-7aa	Acid phosphatase	Lysosomes
TM-24aa-SDYQRLNLKL	TGN38	TGN
TM-79aa-LAYSAF-141aa	Poly-Ig receptor	Basolateral PM

TM = transmembrane domain.
All the references for these signals are reported in ref. 44.

anchor is added after cleavage of the signal sequence and removal of a short hydrophobic C-terminal peptide,[54] that directs its own cleavage and replacement with preassembled GPI-anchor precursor. Like the N-terminal leader sequences, this C-terminal GPI signal peptide does not have a conserved sequence, but does have some characteristic features. Fusion of the GPI signal to the carboxyl terminus of a secreted protein leads to the expression of that protein on the cell surface as a GPI-anchored protein. GPI-anchors localize in membranes rich of glycolipids, sphingolipids and cholesterol and resist solublization by several detergents. However, in contrast with transmembrane proteins, GPI-anchors associate via their acyl chains with the outer cell membrane leaflet, and, as such, are attached more loosely to the plasma membrane. This loose binding leads to the spontaneous transfer of these molecules between membranes of different cells in vitro and in vivo.

Signals for Targeting of Soluble Proteins to Regulated Secretory Granules

It is now widely accepted that the selective aggregation of regulated secretory proteins is part of the mechanism which ensures the efficient segregation of these proteins from constitutive secretory ones.[9] This comes largely from studies on granins (chromogranins/secretogranins).[55] In this case, the selective aggregation in the TGN is induced by specific physicochemical factors (pH, Ca concentration, etc.) (-). To generate a properly assembled secretory granule, the aggregated regulated secretory protein could somehow interact with a membrane component. The precise nature of the signal, or of the structure on regulated secretory proteins that mediate this interaction, referred to as an S-M (Secretory protein-Membrane recognition) signal, remains to be established. In chromogranins such an S-M signal appears to be associated with the 20 amino acid loop formed by the single intramolecular disulfide bond in this protein. The sequence comparison of a wide variety of regulated secretory proteins has not revealed a consensus sequence that qualifies as an S-M signal. Loh and colleagues have reported that the 24 N-terminal residues of pro-opiomelanocortin (POMC), which comprise a 13 amino acid amphipatic loop, stabilized by a disulfide bond, are able to direct a reporter protein to the regulated pathway.[56,57] There is no extensive sequence homology between the chromogranin B and the POMC loops, except for an LS/TXXS motif. It is proposed that the hallmark of the S-M signal of regulated secretory proteins may be a degenerate sequence folding into a common receptor-recognizable structure. Very recently, carboxypeptidase E was identified as a bona fide receptor for this conformational motif and responsible for sorting to granules.[58]

Sorting in the Cytosol

In the absence of the leader sequence for entering the secretory pathway, the nascent polypeptide is translated from free ribosomes in the cytosol. From the cytosol, proteins have access to a number of different intracellular locations. Many motifs have been identified which specify the targeting from the cytosol to different organelles. It should be noted that in some cases (e.g., nucleus, cytoplasmic face of plasma membrane), newly synthesized proteins will have a direct access to the relevant compartment, while in other cases, their translocation through intracellular membranes will be required (for instance, mitochondrial proteins).

Nuclear Targeting

In eukaryotic cells, there is a continuous exchange of macromolecules between the nucleoplasm and the cytosol. The nuclear membrane is made of an external membrane, an extension of the ER membrane, and of an internal membrane, the nuclear lamin network. Proteins enter the nucleus through the 90 Å wide nuclear pores[59-61] and their transport is mediated by shuttle proteins (kariopherins or importins) recognizing nuclear localization sequences (NLS). The nuclear pore complex has a molecular sieve function, whereby molecules smaller than 40-45 kDa can diffuse freely in and out of the nucleus, independent of temperature and energy. Proteins larger than 45 kDa require a nuclear localization sequence (NLS) in order to be targeted specifically to the nucleus. NLS-dependent protein transport is a two-step ATP-dependent process. Transport across nuclear pores is bidirectional (for a review see ref. 62).

The nuclear localization sequences (NLS) are generally short stretches of amino acids (6-15 aa). NLSs are not cleaved during transport[61] and are recurrently required throughout cell division which, in the case of eukaryotic cells, involves nuclear envelope breakdown. NLSs have been identified on the basis of two major criteria: (i) mutation or deletion of the NLS leads to cytoplasmic localization of the protein and (ii) addition to heterologous proteins of the NLS leads to their nuclear targeting. Examples of NLS which have been shown to confer nuclear localization to reporter proteins are listed in Table 5.3.

Table 5.3. Examples of nuclear localization signals

NLS	Protein
PKKKRKV	SV40 T-ag
VSRKRPRP	Polyoma T
KRPAATKKAGQAKKKKLD	Nucleoplasmin
VRTTKGKRKRIDV	Lamin L1
PEEVKKRKKAV	Cofilin
PAAKRVKLD	Human-c-myc
KRPRP	Ad7 E1a
PNKKKRK	SV40VP2/3
PQPKKKPL	Human p53
SALIKKKKKMAP	Mouse c-abl IV
KSRKRKL	v-jun/c-jun
KTRKHRGKHRKHPG	Ribosomal protein L29
NKIPIKDLLNPQVRILESWFAKNIENPYLDT	Yeast MATa2
AAFEDLRVLS	Influenza virus nucleoprotein
GKKRSKAK	Yeast histone 2B
RVTIRTVRVRRPPKGKHRK	Monkey v-sis
RESGKKRKRKRLKPT	Human PDGF A (longer form
TKKQKT	Prothymosin a
EYLSRKGKLEL	VirD2protein (octopine, *Agrobacterium tumefaciens*)

These are NLS that have been shown to be able to target heterologous proteins to the nucleus either as fusion protein derivative or peptide covalently coupled to the carrier (all the references for these NLS are reported in ref. 62).

The first NLS to be characterized and studied is the sequence PKKKRKV derived from the large T antigen of SV40.[63] This sequence, able to confer a nuclear localization to a large number of proteins, has also been used to target antibodies to the nucleus.[64]

No general consensus sequence has been revealed by the analysis of a large number of NLSs, although most of them are characterized by the presence of basic amino acids (K and R). Based on competition studies, proteins possessing these T-antigen-like sequences may belong to a general class of nuclear proteins that reach the nucleus by the same mechanism. The targeting efficiency of NLSs can be modulated by different mechanisms: (i) modification (such as phosphorylation) of the flanking regions; (ii) the presence of multiple copies, especially in the case of a weak NLS and/or in the case of bipartite NLSs, which consist of two stretches of basic residues separated by a spacer, (iii) the distance from each other (length of the spacer). This implies that conformation and hydrophobicity may be important.

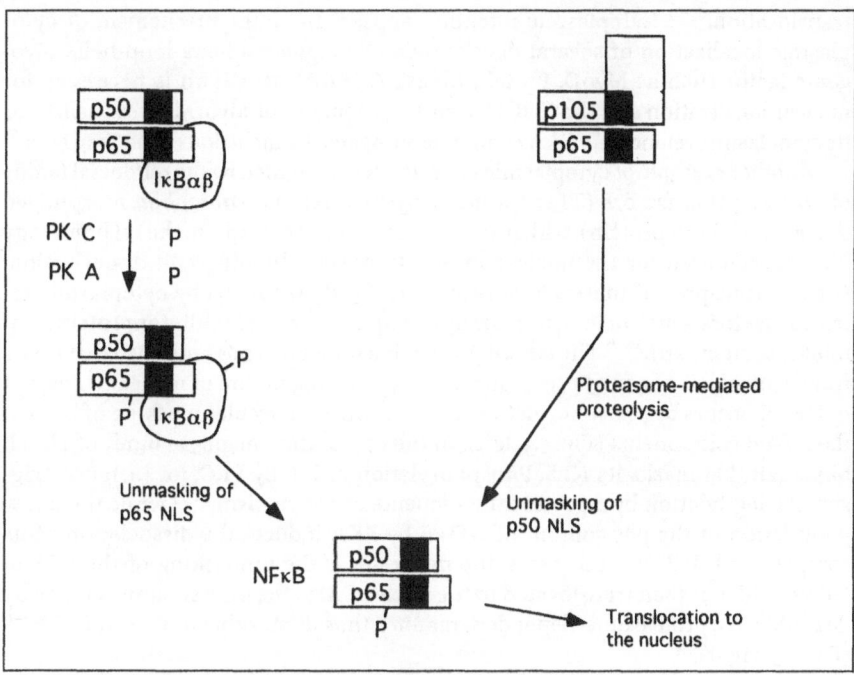

Fig. 5.3. Regulation of NF-κB nuclear transport by phosphorylation and intra-/inter-molecular masking. The NLS is depicted as a black box.

Regulated Nuclear Protein Transport

While many proteins are constitutively targeted to the nucleus, many others are only conditionally targeted to the nucleus, being preferentially cytoplasmic. These proteins are targeted to the nucleus by regulatory mechanisms that modulate NLS accessibility, such as NLS masking and phosphorylation enhancing NLS-dependent nuclear transport. Examples of regulated nuclear transport concern, mostly, transcription factors, but also kinases, cyclins and viral proteins. Mechanisms have been discovered, whereby not only the amount of a given protein targeted to the nucleus can be modulated, but also its rate of transport according to the cell cycle, the metabolic state or in response to incoming extracellular signals (for a review see ref. 62).

Cytoplasmic retention is one of the mechanisms of regulated nuclear import and implies the presence of a cytoplasmically localized "anchor" protein or retention factor which specifically binds and mask the NLS of the nuclear protein. An example of cytoplasmic retention has been described for the glucocorticoid receptor which, in the absence of the hormone, forms a complex with the HSP90 heat shock protein masking its NLS. Hormone binding to the receptor induces the hormone-receptor complex translocation to the nucleus, binding to specific DNA sequences and modulation of transcription. The hormone binding domain of several steroid receptor proteins share the same properties and, if fused to other reporter proteins such as MyoD or myc, confer to them the hormone inducible nucleus translocation.[65-67] Cytoplasmic retention appears to be the mechanism of cytoplasmic localization of several developmentally regulated helix-loop-helix myogenic factor such as MyoD. Protein kinase A (PKA) activation is necessary for nuclear localization and the COOH-terminal sequence of MyoD is responsible for its cytoplasmic retention, which cannot be abrogated by inclusion of the T-ag NLS.[68]

Another example of cytoplasmic retention is represented by the rel/dorsal family of transcription factors (TFs) (protooncogene c-rel, the *Drosophila* morphogen dorsal and NF-κB protein), which share a 300 amino acid region, the rel homology domain, required for the nuclear localization, DNA binding and dimerization. Nuclear transport of these TFs appears specifically regulated by cytoplasmic retention factors which belong to a family of sequence-related inhibitor proteins (inhibitor protein IκB).[69-71] The NF-κB protein is a dimer of a p65 and a p50 that arises from proteolysis of a p105 precursor (Fig. 5.3). The regulation of nuclear transport of NF-κB occurs by phosphorylation and intra/intermolecular masking of both of the NF-κB components (Fig. 5.3, left): in the cytoplasm, the p65 subunit of NF-κB binds IκB that masks its NLS. Phosphorylation of IκB by PKC, for instance, triggers its degradation by a proteasome-dependent mechanism,[72-73] while the phosphorylation of the p65 subunit of NF-κB by PKA induces the dissociation of its complex with IκB. In both cases, the net result is the unmasking of the NLS of NF-κB, which is then translocated to the nucleus. Also the p105 subunit is partially degraded in a proteasome-dependent manner, thus unmasking its C-terminal NLS (Fig. 5.3, right).[74]

In many instances, such as for NF-κB, the phosphorylation of residues in the proximity of the NLS positively modulates the accessibility of the NLS, otherwise "invisible", thus modulating the translocation process. These phosphorylation sites may be classified according to the positive or negative influence of the phosphorylation event on nuclear translocation.

For instance, casein kinase II increases the $t_{1/2}$ of nuclear import of T-antigen by phosphorylating a serine 13 residues N-terminal to the NLS.[75] On the other hand, phosphorylation by cdc2 of a Thr 2 residues upstream of the NLS of T-antigen inhibits its nuclear accumulation by allowing interaction with a cytoplasmic retention factor.

A variety of different signaling events related to cell cycle, extracellular ligands, metabolic state, proliferative and differentiative stage of the cell can affect the activity of numerous kinases which, in turn, will modulate the localization of nuclear targeted proteins.[76,77]

Nucleolar Localization Signals

Nucleolar localization sequences (NOSs) appear to be related to NLS in that they consist of short sequences of basic amino acids. Both NLSs and NOSs should be present in a protein to be targeted to the nucleolus.

In the case of the *Xenopus* nucleolar protein No38, this protein enters the nucleus by a NLS and in the nucleolus by oligomerization.

NOS signals of HSP70 (FKRKHKKDISQNKRAVRR), human HIV Tat (GRKKRRQRRRAP) and the HTLV-1 Rex protein (MPKTRRRPRRRSQRKRPPTP) are able to direct reporter proteins to the nucleolus.[78,79]

Nuclear Export

As discussed, the nuclear pore is a bidirectional trafficking barrier. Recently, it has been reported that similarly to the nuclear import signals (NLS), specific sequences can determine the selective nuclear export of proteins. Only a few of these nuclear export signals (NESs) have been so far characterized. Two of these sequences, the NES of the Protein kinase inhibitor PKI (a PKA specific inhibitor protein) (LALKLAGLNIN)[80] and the NES of the HIV-1 rev protein (LPPLGRLTLN),[81] have been shown to be necessary and sufficient when mutagenized into heterologous proteins to work as export signals.[82,83]

Viral Transport into the Nucleus

The nuclear transport of viral nucleic-acid/protein complexes is of fundamental importance for viral replication. For enveloped viruses, after the viral replication occurs in the nucleus, the process of maturation occurs in the cytoplasm; therefore, mechanisms for nuclear export are of great importance as well. Different viruses use different strategies.[84-87] The problem of nuclear transport is directly related to the capacity of some viruses to infect postmitotic cells. Indeed, many oncoretroviruses only infect dividing cells because the preintegration complex can only access the nucleus upon disruption of the nuclear membrane during cell division. Lentiviruses, such as HIV-1, can support the nuclear targeting of the preintegration complex and hence can infect nondividing cells. Import is mediated by viral proteins as well as by the action of cellular kinases.

The preintegration complex contains the viral nucleic acid, the reverse transcriptase, the integrase and two other viral proteins: the Gag matrix associated protein MA and the accessory gene product Vpr.[88-90] These two proteins play a role in the cytoplasm nuclear traffic of the afferent and the efferent phase of viral infection. The protein MA contains a NLS that is also active when fused to heterologous proteins.[91] Peptides containing the NLS of T-antigen or of MA inhibit the nuclear targeting and the infectivity of the virus. MA protein contains two competing targeting signals, one, myrystylation-dependent for targeting the protein to the cytoplasmic face of the plasma membrane, the second one for targeting it to the nucleus.[92,93] The NLS of MA is activated by phosphorylation. The nonphosphorylated protein is localized to the membrane, while the phosphorylated protein is found in the preintegration complex. During the viral packaging, MA is further phosphorylated by a cellular kinase that is coassembled with the virus. This additional phosphorylation favors the disassembly of MA from the membrane.

This provides a very good example of conflicting and alternative signals present on the same protein, with their activity being modulated by the cellular context.

Targeting of Cytosolic Proteins to the Cytoplasmic Face of Plasma Membrane

Anchor sequences which specifically fix cytoplasmic proteins to the inner face of plasma membranes have been described in proteins involved in receptor-mediated signal transduction by growth factors and hormones. One example of this type of motif is a C-terminal 4 amino acid peptide, CAAX, that was found responsible for targeting ras proteins to the cytoplasmic face of plasma membrane.[94] The CAAX box acts together with a second signal to target plasma membrane localization.[95] This localization involves a complex series of post-translational modifications which comprise polyisoprenilation, C-terminal proteolysis, carboxyl-methylation and palmoytilation. Mutation of Cys 186, which is also the palmitoylation site, blocks both membrane localization and transformation. A 20 aa C-terminal sequence of p21-ras, encompassing both signals, has been used to redirect and anchor heterologous cytoplasmic proteins to the cytoplasmic face of plasma membrane.[96] Another example of this sort is provided by the 15 N-terminal residues of the pp60src protein kinase, which are a signal for the myrystylation of the protein and its localization at the plasma membrane.[97]

Mitochondrial Targeting

The majority of mitochondrial proteins are encoded in the nucleus, synthesized in the cytoplasm and then targeted to mitochondria. The remaining few mitochondrial proteins are encoded by the mitochondrial genome. The synthesis of mitochondrial proteins and their import from the cytoplasm are tightly coupled and, at least in some cases, a cotranslational import reaction seems to occur (for a review see refs. 98, 99). Most nuclear-encoded mitochondrial proteins contain N-terminal extensions of the protein (presequences), the so-called mitochondrial targeting signals (MTSs) which are in several cases, sufficient to transport attached foreign 'passenger' proteins to and across mitochondria.[100,101] As is the case for other targeting signals, there is no significant sequence homology among the known MTSs. In this case, the information of the presequences is not determined by their primary structure, although these sequences show a preponderance of basic, hydrophobic and hydroxyl-carrying residues and lack acidic residues. On the contrary, several studies based on known and artificial presequences, consisting most exclusively of leucine, arginine and serine, have shown their potential to form amphipathic, positively charged α-helices which appear to be crucial for transport.[102-104] Proteins must be unfolded for translocation, a state which is maintained in the cytoplasm by a variety of chaperons, until the complex binds to the translocating proteins on the trans side of the membrane.[105]

Although most of the mitochondrial proteins contain N-terminus presequences, the bcl-2 oncoprotein contains a C-terminus signal anchor sequence which has been shown to be necessary and sufficient to mediate mitochondrial insertion.[106] The product of the human protoncogene bcl-2 is an integral membrane protein that is located in the perinuclear endoplasmic reticulum, the nuclear envelope and the mitochondria.[107] The transmembrane segment located at the C-terminus of the molecule is responsible for targeting and insertion of bcl-2, and of other reporter proteins, into the outer membrane of mitochondria in a N_{cyto}-C_{in} orientation; deletion of this sequence abolishes the mitochondrial location converting bcl-2 to a cytosolic protein.[108]

Peroxisomal Targeting

All peroxisomal proteins are synthesized on cytosolic free polyribosomes, but the entry into peroxisomes does not follow the successive steps required for the import of proteins from the cytoplasm into ER or mitochondria, which are recognition, translocation and maturation. Proteins must be unfolded for translocation, a state that needs molecular chaperones to be maintained, and, after translocation, be folded on the internal side of the membrane. In the case of peroxisomes, proteins are imported post-translationally to the organelles and may be translocated into the peroxisomal matrix in a folded or oligomerized state.[109,110] In contrast to mitochondrial proteins translocation, cleavage of the presequences is not a necessary step in import of peroxisomal proteins. However, the N-terminal presequences of mammalian thiolases, necessary for the correct targeting, are cleaved.[111] Several signals direct protein traffic to peroxisomes. Most peroxisomal proteins contain a tripeptide (Ser-Lys-Leu) present at the COOH-terminal, the so-called peroxisomal targeting signal of type 1 (PTS1) which was described for the first time in firefly luciferase.[112] This short carboxy terminal region, present in many peroxisomal proteins,[113] is sufficient to target heterologous proteins to peroxisomes. Recently, a novel COOH-terminal peroxisomal targeting sequence have been identified in human catalase.[114] Addition of the last four amino acids (-KANL) of the enzyme at the C-terminus of a reporter protein, are necessary and sufficient to direct it to peroxisomes both in human fibroblast and in *Saccharomyces cerevisiae*.

Peroxisomal targeting sequence of type 2, composed of a motif at the N-terminus, was found in some peroxisomal-matrix proteins. It was first found in the β-oxidation enzyme 3-keto-acyl-CoA thiolase.[115] The sequence is usually composed of a conserved nonapeptide with several common amino acids $(R/K)(L/V/I)(X_5)(H/Q)(L/A)$.[116]

A peroxisomal membrane protein targeting signal has been described for PmP47 from the yeast *Candida boidinii*.[117]

Signals for Rapid Degradation of Cytosolic Proteins

Signals for degradation constitute a special class of targeting signals. Multiple pathways of protein degradation for cytosolic proteins operate within the cells and numerous amino acid sequences responsible for the high turnover of cytoplasmic and nuclear proteins have been characterized, including those for ubiquitination. Regulated proteolysis is part of the quality control system of the cell to remove misfolded proteins, but is also a regulatory system that allows the levels of a given protein to be rapidly and irreversibly controlled. It is clear that the specificity of the system is very important as to avoid the danger of aspecific proteolysis.

Lysosomal Degradation of Cytosolic Proteins

A selective protein import pathway that appears to be similar to the membrane translocation systems of the ER and mitochondrial membranes, exists for the up-take and degradation by lysosomes of some cytosolic proteins, such as RNAse A. The degradation of RNAse A by lysosomes is increased upon growth factor removal. The sequence KFERQ (aa 7-11 of RNAse A) is essential for this degradation.[118,119] The exact sequence is only present in proteins of the RNAse A family, but many other proteins contain sequences of similar charge and hybrophobicity. All these proteins bind prp73, a heat shock protein of the hsp70 family.[119,120] This protein also interacts with clathrin, inducing its disassembly. Clathrin contains two KFERQ-like peptides, and is thought to play an important role in the selection of cytosolic proteins to be degraded by lysosomes.

A receptor mediating the selective uptake of cytosolic proteins containing the KFERQ motif into lysosomes has recently been identified, the lysosomial glyco-protein LGP96. The substrates of this receptor appear to recognize the 12 residue cytosolic tail of this receptor before being transported into lysosomes for degradation.[121] Overexpression of human LGP96 in Chinese hamster ovary cells increases the activity of the selective lysosomal proteolytic pathway, both in vivo and in vitro.

The Ubiquitin-Proteasome Pathway

The ubiquitin-proteasome pathway represents a specific degradation system for cytosolic proteins (for reviews see refs. 122-126). Proteasomes are multienzymatic complexes of approximately 2000-kDa, that degrade selectively, with an ATP-dependent mechanism, poly-ubiquitinated proteins. Ubiquitination is thus a way for adding, post-translationally, a targeting tag for degradation. 26S proteasome complexes are made of the 20S proteasome core and of the 19S caps, which contain the ATP binding sites and are probably involved in unfolding of the incoming proteins. The mechanism of poly-ubiquitination requires the action of three classes of proteins E1, E2 and E3, involved in the recognition, modification and transport of the substrate to be degraded in proteasomes.

The enzyme E1 utilizes the energy deriving from ATP hydrolysis to activate ubiquitin, forming an intermediate high-energy tioesther bond with ubiquitin. The enzyme E2 catalyses the transfer of activated ubiquitin to the substrate, which is recognized by E3. The E3 ligase closes the cycle, catalyzing isopeptide bond formation between ubiquitin and the substrate. Additional ubiquitin molecules are usually added to the substrate in a processive way by the same enzyme cascade, to form a chain or chains of ubiquitin molecules (poly-ubiquitin). Poly-ubiquitinated proteins are targets for destruction by the 26S proteasome.

As E3 enzymes specifically bind protein substrates, they play an important role in the selection of proteins for degradation and, as such, may represent good targets for antibody-mediated degradation engineering (see chapter 7).

The ubiquitin-proteasome pathway is involved in many cellular functions, including the degradation of short-lived regulatory proteins (oncogenes, transcription factors, cyclins) and antigen presentation of viral antigens by Class I MHC complex. Some of the signals recognized by the proteins of the multisubunit proteasome complex have been identified by studying the sequence requirements for short-lived proteins. Some of these degradation signals can be considered to be autonomous and dominant, in that they are able to confer the property of rapid turnover to reporter proteins. Following are some examples of signals for degradation that have been recently identified.

One important mechanism is commonly referred to as the N-end rule. The N-end rule can be defined as the relation between the metabolic stability of a protein and the identity of its N-terminal residues.[127,128] Some amino acids function as destabilizing N-terminal residues, and, by mutating the N-terminal residues, the half life of a protein can be experimentally modulated. This pathway has been described in all organisms examined, from bacteria to mammals. The destabilizing activity of N-terminal residues requires their physical binding by the E3 protein of the ubiquitination complex, which then induces the polyubiquitination of an internal lysine residue of the target protein. In eukaryotes, amino acids have been classified according to their capacity to be recognized by the system. Among these, type 1 binding site of E3 binds N-terminal basic residues Arg, Lys, or His, whereas the type 2 site binds N-terminal hydrophobic amino acids Phe, Leu, Trp, Tyr or Ile. Conversely, a residue that is not recognized by the E3 component is a stabilizing N-terminus residue, according to the N-end rule.

Another example of an ATP-dependent degradation mediated by proteasomes, but not utilizing ubiquitination, is the proteolysis of ornithine decarboxylase (ODC). ODC, the key enzyme in the biosynthesis of polyamines, is a short-lived protein degraded in a highly regulated manner. Regulated degradation of ODC requires association with another small protein, called antizyme. Polyamines regulate ODC activity via induction of antizyme, which binds to ODC and promotes its degradation. Two domains within ODC are necessary for polyamine-dependent degradation, the antizyme binding domain near the N-terminus, and the C-terminus degradation domain. The mechanism by which antizyme promotes degradation, in the absence of ubiquitin, is still not elucidated. However, the antizyme contains a N-terminus domain which can be grafted to heterologous proteins to make them labile.[129]

The human papilloma virus oncoprotein E6 acts as an E3-type enzyme, in that it targets the cellular p53 protein for degradation. E6-p53 complex formation requires the cellular protein E6-AP (E6 associated protein),[130] which undergoes specific ubiquitinylation. The modified E6-AP then acts as an E3 enzyme which ubiquitinylates the associated p53, leading to its instant degradation. The viral E6 protein, therefore, facilitates the recognition of the p53 substrate by cellular ubiquitin-conjugating enzymes.

Ubiquitin itself can be considered as a degradation signal if fused to heterologous proteins.[131] The fusion protein ubiquitin-proline-β-gal has a very short half life in yeast. The demonstration that a noncleavable mono-ubiquitin fusion may function as a degradation signal by ubiquitin-dependent proteolysis, may be applicable to other proteins, possibly exploiting the fact that a functional ubiquitin molecule can be reconstituted from two split halves by a heterologous protein-protein interaction.[128]

Cyclin A and B are targeted for ubiquitination during anaphase, by a nine amino acid conserved motif known as the "destruction box".[132,133] Cyclin mutants lacking the destruction box are neither ubiquitinated nor degraded, and arrest the cell cycle in late anaphase.[134,135] A fusion of protein A, with the destruction box from cyclin B can be ubiquitinated, although in some cases, the destruction box is not a sufficient signal to direct ubiquitination.

PEST Sequence and Proteolysis

PEST regions were first identified on the basis of a correlation between metabolic stability and some protein sequences. Some of the rapidly degraded proteins (as myc, fos, jun, p53) contain hydrophilic stretches of about 12 amino acids enriched in proline (P), glutamate (E), serine (S) and threonine (T).[136,137] These regions appear to represent constitutive targeting signals for rapid degradation as in the case of mouse ODC,[138] or conditional targeting signals as for Cyclic-AMP dependent kinase.[139] PEST regions are transplantable proteolytic signals,[138-140] although they are not proven to be essential nor sufficient for targeting proteins for destruction. The mechanism by which PEST sequences act has not been elucidated yet, but the involvement of proteasome degradation for some of these proteins has been suggested.[137]

mRNA Targeting

One further way that proteins have to reach their subcellular locations is to localize the messenger RNA that encodes them. Since mRNAs act as the templates for translation, their localization allows specific protein to be synthesized in the subcellular regions where they are required and prevents their expression in regions where they are not. In principle, localized protein synthesis would seem to be a very efficient way to target proteins to the correct sites as presumably more energy is needed to localize many protein molecules than a single mRNA that can be translated multiple times. In addition to cutting intracellular transport costs, localized translation opens up the possibility of local translational control.

mRNA Targeting in Polarized Cells

The generation of a polarized cell type, such a neuron, requires the asymmetric distribution of many cellular proteins. Many examples of discrete localization of specific mRNAs have been reported, providing a general mechanism for protein targeting in polarized cells. For example, the mRNAs encoding the cytoskeletal protein MAP2 and the α subunit of Ca/calmodulin dependent protein kinase II have been found to localize to dendrites, the regions of the neurons where most synaptic inputs are received, while the mRNA for the microtubule associated protein τ is localized to axons.[141] Furthermore, it is also known that protein synthesis occurs in dendrites and that polyribosomes are enriched beneath postsynaptic sites. Thus, it has been proposed that translation might be regulated independently at each of the many postsynaptic sites within a single neuron, and that this may play a role in activity dependent synaptic plasticity.

Another reason why a specific mRNA might be localized is to prevent the expression of the protein it encodes in the wrong region of the cells. This seems to be the case for myelin basic protein (MBP), a component of the myelin sheath that oligodendrocytes wrap around axons.[142]

mRNA Targeting to Establish Polarity

Not only does mRNA localization serve to deliver proteins to the appropriate sites within a polarized cell, it can also play a direct role in the establishment of polarity. This is particularly well studied in the *Drosophila* oocyte, where more than ten mRNAs have now been found to localize to one of three distinct positions within this single cell, and these transcripts play a variety of roles in the specification of the embryonic body plan. Thus, mRNA localization is often tightly coupled to translational control. If it is important for a cell to synthesize a protein in a particular place, then the translation of the mRNA must be repressed until it is localized and the localization of the mRNA may then lead to the local production or even secretion of the corresponding protein. Indeed, there are already several examples where the direct linkage between the translational control and localization has been demonstrated (for a review see ref. 143).

mRNA Localization Signals

A common feature of all the localized mRNAs examined so far is that cis-acting sequences required for localization reside in the 3'UTRs of the transcript. In the cases where these sequences have been precisely mapped, they have turned out to be relatively large. This is probably because these signals will contain multiple protein-binding sites, since localization often involves several steps, each requiring different trans-acting factors. So far, four distinct mechanisms for mRNA localization have been identified: spatial control of mRNA stability, anchoring to localized binding sites, vectorial nuclear export and directed transport.

Examples of mRNA localization signals that have been mapped within the 3'UTR of the corresponding mRNAs include those of bicoid[145] (400 nucleotides), and VG1[146] (340 nucleotides).

Conclusions

For some intracellular compartments it is now relatively easy to exploit, among the identified signals, those that are most suitable for the targeting of an antibody. The choice must keep into account the following considerations: (i) the signal should be relatively short; (ii) it should confer the desired targeting also in the context of an antibody protein; (iii) it should not harbor cryptic degradation signals, unless these are specifically engineered and (iv) it should not interfere with the folding and the antigen binding of the antibody. For other intracellular compartments, where targeting signals have not been pinpointed to a small peptide sequence yet, the appropriate desired targeting may nonetheless be achieved by constructing a C-terminal fusion of the antibody domain with a protein domain targeted there.

References

1. Blobel G, Sabatini DD. Ribosome-membrane interaction in eukaryotic cells. In: Manson LA, ed. Biomembranes. New York: Plenum, 1971; 2:193-195.
2. Milstein C, Brownlee GG, Harrison TM et al. A possible precursor of immunoglobulin light chain. Nature New Biol 1972; 239:117-120.
3. Blobel G, Dobberstein B. Transfer of proteins across membranes. Presence of proteolytically processed and unprocessed nascent immunoglobulin light chains on membrane-bound ribosomes of murine myeloma. J Cell Biol 1975; 67:835-851.
4. Luzio JP, Banting G. Eukaryotic membrane traffic: retrieval and retention to achieve organelle residence. TIBS 1993; 18:395-399.
5. Trowbridge IS, Collawn JF, Hopkins CR. Signal-dependent membrane protein trafficking in the endocytic pathway. Ann Rev Cell Biol 1993; 9:129-161.
6. Rothmann JE, Orci L. Molecular dissection of the secretory pathway. Nature 1992; 355:409-415.
7. Rothman J, Wieland FT. Protein sorting by transport vesicles. Science 1996; 272:227-234.
8. Machamer CE. ER-Golgi membrane traffic and protein targeting. In: Hurtley SM, ed. Protein Targeting. IRL Press. 1996; 123-151.
9. Ohashi M, Bauerfeind R, Huttner WB. Biogenesis of constitutive secretory vesicles, secretory granules and synaptic vesicles—facts and concepts. In: Hurtley SM, ed. Protein Targeting. Series Editors: Hames BD, Glover DM. IRL Press. 1996:152-175.
10. Wieland FT, Gleason ML, Serafini T et al. The rate of bulk flow from the endoplasmic reticulum to the cell surface. Cell 1987; 50:289-300.
11. Pfeffer SR, Rothman JE. Biosynthetic protein transport and sorting by the endoplasmic reticulum and Golgi. Ann Rev Biochem 1987; 56:829-852.
12. Griffiths G, Doms RW, Mayhew T et al. The bulk-flow hypothesis: not quite the end. Trends Cell Biol 1995; 5:9-13.
13. Singer SJ. It's important to concentrate. Trends Cell Biol 1995; 5:14-15.
14. Balch WE, Farquhar MG. Beyond bulk flow. Trends Cell Biol 1995; 5:16-19.
15. Aridor M, and Balch WE. Principles of selective transport: coat complexes hold the key. Trends Cell Biol 1996; 6:315-320.
16. Hammond C, Helenius A. Quality control in the secretory pathway. Curr Opin Cell Biol 1995; 7:523-529.
17. Bonifacino JS, Cosson P, Klausner RD. Colocalized transmembrane determinants for ER degradation and subunit assembly explain the intracellular fate of TCR chains. Cell 1990; 63:503-513.
18. Klausner RD, Sitia R. Protein degradation in the endoplasmic reticulum. Cell 1990; 62:611-614.

19. Bonifacino JS, Cosson P, Shah N et al. Role of potentially charged transmembrane residues in targeting proteins for retention and degradation within the endoplasmic reticulum. EMBO J 1991; 10:2783-2793.

20. Letourneur F, Hennecke S, Demolliere C et al. Steric masking of a dilysine endoplasmic reticulum retention motif during assembly of the human high affinity receptor for immunoglobulin. E J Cell Biol 1995; 129:971-978.

21. Lord JM. Go outside and see the proteasome. Curr Biol 1996; 6:1067-1069.

22. Ward CL, Omura S, Kopito RR. Degradation of CFTR by the ubiquitin-proteasome pathway. Cell 1995; 83:121-127.

23. Jensen TJ, Loo MA, Pind S et al. Multiple proteolytic pathways, including the proteasome, contribute to CFTR processing. Cell 1995; 83:129-135.

24. von Figura K. Molecular recognition and targeting of lysosomal proteins. Curr Opin Cell Biol 1991; 3:642-646.

25. Ponnambalam S, Banting G. Protein secretion: Sorting sweet sorting. Curr Biol 1996; 6:1076-1078.

26. Gahmberg CG, Tolvanen M. Why mammalian cell surface proteins are glycoproteins? TIBS 1996; 21:308-311.

27. Scheiffele P, Peranen J, Simons K. N-glycans as apical sorting signals in epithelial cells. Nature 1995; 378:96-98.

28. Munro S, Pelham HRB. A C-terminal signal prevents secretion of luminal ER proteins. Cell 1987; 48:899-907.

29. Pelham HRB. Control of protein exit from the endoplasmic reticulum. Ann Rev Cell Biol 1989; 5:1-23.

30. Pelham HRB. The retention signal for soluble proteins of the endoplasmic reticulum. TIBS 1990; 15:483-485.

31. Nilsson T, Warren G. Retention and retrieval in the endoplasmic reticulum and the Golgi apparatus. Cur Opin Cell Biol 1994; 6:517-521.

32. Biocca S, Tafani M, Cattaneo A. Assembled IgG molecules are exported from the endoplasmic reticulum in myeloma cells despite the retention signal SEKDEL. FEBS letters submitted.

33. Yoshimori T, Semba T, Takemoto H et al. Protein disulfide-isomerase in rat exocrine pancreatic cells is exported from the endoplasmic reticulum despite possessing the retention signal. J Biol Chem 1990; 265:15984-15990.

34. Sitia R, Neuberger M, Alberini C et al. Developmental regulation of IgM secretion: the role of the carboxy-terminal cysteine. Cell 1990; 60:781-790.

35. Kerem A, Kronman C, Bar-Nun S et al. Interrelations between assembly and secretion of recombinant human acetylcholinesterase. J Biol Chem 1993; 268:180-184.

36. Jackson MR, Nilsson T, Peterson PA Identification of a consensus motif for retention of transmembrane proteins in the endoplasmic reticulum. EMBO J 1990; 9:3153-3162.

37. Shutze MP, Peterson PA, Jackson MR. An N-terminal double-arginine motif maintains type II membrane proteins in the endoplasmic reticulum. EMBO J 1994; 13:1696-1705.

38. Shaper JH, Shaper NL. Enzymes associated with glycosylation. Curr Opin Cell Biol 1992; 2:701-.

39. Machamer CE. Targeting and retention of Golgi membrane proteins. Curr Opin Cell Biol 1993; 5:606-612.

40. Munro S. An investigation of the role of transmembrane domains in Golgi protein retention. EMBO J 1995; 14:4695-4704.

41. Williams GT, Ventikarman AR, Gilmore DJ et al. The sequence of the μ transmembrane segment determines the tissue specificity of the transport of immunoglobulin M to the cell surface. J Exp Med 1990; 171:947-952.

42. Humphrey JS, Peters PJ, Yuan LC et al. Localization of TNG38 to the trans-Golgi network: involvement of a cytoplasmic tyrosine containing motifs. J Cell Biol 1993; 120:1123-1135.

43. Stanley KK, Howell KE. TNG38/41: a molecule on the move. Trends Cell Biol 1993; 3:252-.

44. Marks MS, Ohno H, Kirchhausen T et al. Protein sorting by tyrosine-based signals: adapting to the Ys and wherefores. Trends Cell Biol 1997; 7:124-128.

45. Trowbridge IS. Endocytosis and signals for internalization. Curr Opin Cell Biol 1991; 3:634-641.

46. Jones BG et al. Intracellular trafficking of furin is modulated by the phosphorylation state of a casein kinase II site in its cytoplasmic tail. EMBO J 1995; 14:5869-5883.

47. Seethaler G, Tooze S, Shields D. Fusion and confusion in the secretory pathway. Trends Cell Biol 1996; 6:239-242.

48. Peters C, Braun M, Weber B et al. Targeting of a lysosomal membrane protein: a tyrosine-containing endocytosis signal in the cytoplasmic tail of lysosomal acid phosphatase is necessary and sufficient for targeting to lysosomes. EMBO J 1990; 9:3497-3506.

49. Sandoval IV, Bakke O. Targeting of membrane proteins to endosomes and lysosomes. Trends Cell Biol 1994; 4:292-297.

50. Letourneur F, Klausner R. A novel Di-leucine motif and a tyrosine-based motif independently mediate lysosomal targeting and endocytosis of CD3 chains. Cell 1992; 69:1143-1157.

51. Roher J, Schweizer A, Russel et al. The targeting of Lamp1 to lysosomes is dependent on the spacing of its cytoplasmic tail tyrosine sorting motif relative to the membrane. J Cell Biol 1996; 132:565-576.

52. Ilangumaran S, Robinson PJ, Hoessli DC. Transfer of exogenous glycosylphosphatidylinositol (GPI)-linked molecules to plasma membranes. Trends Cell Biol 1996; 6:163-167.

53. Udenfriend S, Kodukula K. How glycosylphosphatidylinositol-anchored membrane proteins are made. Annu Rev Biochem 1995; 64:563-591.

54. Takeda J, Kinoshita T. GPI-anchor biosynthesis. TIBS 1995; 20:367-371.

55. Chanat E, Weiss U, Huttner WB et al. Reduction of the disulfide bond of chromogranin B (secretogranin 1) in the trans-Golgi network causes its missorting to the constitutive secretory pathway. EMBO J 1993; 12:2159-2168.

56. Tam WWH, Andreasson KI, Peng Loh Y. The amino-terminal sequence of proopiomelanocortin directs intracellular targeting to the regulated secretory pathway. Eur J Cell Biol 1993; 62:294-306.

57. Cool DR, Fenger M, Snell CR et al. Identification of the sorting signal motif within pro-opiomelanocortin for the regulated secretory pathway. J Biol Chem 1995; 270:8723-8729.

58. Cool DR, Normant E, Shen F et al. Carboxypeptidase E is a regulated secretory pathway sorting receptor: genetic obliteration leads to endocrine disorders in Cpe(fat) mice. Cell 1997; 88:73-83. (Loh YP figura come ultimo nome il resto coicide exactly).

59. Hannover JA. The nuclear pore at the crossroads. FASEB J 1992; 6:2288-2295.

60. Stochaj U, Silver PA. Nucleocytoplasmic traffic of proteins. Eur J Cell Biol 1993; 59:1.

61. Agutter PS, Prochnow D. Nucleocytosplamic transport. Biochem J 1994; 300:609-618.

62. Jans DA, Hubner S. Regulation of protein transport to the nucleus: central role of phosphorylation. Physiol Rev 1996; 76:651-685.
63. Kalderon D, Roberts BL, Richardson WD et al. A short amino acid sequence able to specify nuclear localization. Cell 1984; 39:499-509.
64. Biocca S, Neuberger M, Cattaneo A. Expression and targeting of intracellular antibodies in mammalian cells. EMBO J 1990; 9:101-108.
65. Hollemberg SM, Cheng PF, Weintraub H. Use of a conditional MyoD transcription factor in studies of MyoD trans-activation and muscle determination. Proc Natl Acad Sci USA 1993; 90:8028-8032.
66. Ellers M, Picard D, Yamamoto KR et al. Chimeras of Myc oncoprotein and steroid receptors cause hormone-dependent transformation of cells. Nature 1989; 340:66-68.
67. Picard D, Salser SJ, Yamamoto KR. A movable and regulable inactivation function within the steroid binding domain of the glucocorticoid receptor. Cell 1988; 54:1073-1080.
68. Rupp RAW, Snider L, Weintraub H. *Xenopus* embryos regulate the nuclear localization of XMyoD. Genes Dev 1994; 8:1311-1323.
69. Henkel T, Zabel U, vanZee K et al. Intramolecular masking of the nuclear localization sequence and dimerization domain in the precursor for the p50 NF-κB subunit. Cell 1992; 68:1121-1133.
70. Zabel U, Henkel T, dos Santos Silva M et al. Nuclear uptake control of NF-κB by MAD-3, an IκB protein present in the nucleus. EMBO J 1993; 12:201-211.
71. Beg AA, Ruben SM, Scheinman RI et al. IκB interacts with the nuclear localization sequences of the subunits of the NF-κB: a mechanism for cytoplasmic retention. Genes Dev 1992; 6:1899-1913.
72. Palombella VJ, Rando OJ, Goldberg AL et al. The ubiquitin-proteasome pathway is required for processing the NF-κB1 precursor protein and the activation of NF-κB. Cell 1994; 78:773-785.
73. Traeckner EB, Wilk S, Baeuerle PA. A proteasome inhibitor prevents activation of NF-κB and stabilizes a newly phosphorylated form of IκB-a that is still bound to NF-κB. EMBO J 1994; 13:5433-5441.
74. Blank V, Kourilski P, Israel A. Cytoplasmic retention, DNA binding and processing of thr NF-κB p50 precursor are controlled by a small region in its C-terminus. EMBO J 1991; 10:4159-4167.
75. Jans DA, Jans P. Negative charge at the casein kinase II site flanking the nuclear localization signal of the SV40 large T antigen is mechanistically important for enhanced nuclear import. Oncogene 1994; 9:2961-2968.
76. Vandromme M, Gauthier-Rouviere C, Lamb N et al. Regulation of transcription factor localization: fine-tuning of gene expression. TIBS 1996; 21:59-64.
77. Faux MC, Scott JD. More on target with protein phosphorylation: conferring specificity by location. TIBS 1996; 21:312-315.
78. Dang CV, Lee WMF. Nuclear and nucleolar targeting sequence of erb-A, c-myb, N-myc, p53, HSP70 and HIV tat proteins. J Biol Chem 1989; 264:18019-18023.
79. Li H, Bingham PM. Arginine/serine-rich domains of the su(wᵃ) and tra RNA processing regulators proteins to a subnuclear compartment implicated in splicing. Cell 1991; 67:335-342.
80. Wen W, Meinkoth JL, Tsien RY et al. Identification of a signal for rapid export of proteins from the nucleus. Cell 1995; 82:463-473.
81. Fischer U, Huber J, Boelens WC et al. The HIV-1 Rev activation domain is a nuclear export signal that accesses an export pathway used by specific cellular RNAs. Cell 1995; 82:475-483.

82. Gerace L. Nuclear exports signals and the fast track to the cytoplasm. Cell 1995; 82:341-344.
83. Fisher U, Michael WM, Luehrmann R et al. Signal-mediated nuclear export pathways of proteins and RNAs. Trends Cell Biol 1996; 6:290-293.
84. Goldfarb D. Viruses: the Trojan horses of the cell. Trends Cell Biol 1996; 6:8.
85. Stevenson M. Portas of entry: uncovering HIV nuclear transport pathways. Trends Cell Biol 1996; 6:9-15.
86. Greber UF, Kasamatsu H. Nuclear targeting of SV40 and adenovirus. Trends Cell Biol 1996; 6:189-195.
87. Whittaker G, Bui M, Helenius A. The role of nuclear import and export in influenza virus infection. Trends Cell Biol 1996; 6:67-71.
88. Karageorgos L, Li P, Burrell C. Characterization of HIV replication complexes early after cell to cell infection. AIDS Res Hum Retroviruses 1993; 9:817-823.
89. Bukrinsky MI, Sharova N, McDonald TL et al. Association of integrase, matrix, and reverse transcriptase antigens of human immunodeficiency virus type 1 with viral nucleic acids following acute infection. Proc Natl Acad Sci USA 1993; 90:6125-6129.
90. Heinzinger NK, Bukrinsky MI, Haggerty SA et al. The Vpr protein of human immunodeficiency virus type 1 influences nuclear localization of viral nucleid acids in nondividing host cells. Proc Natl Acad Sci USA 1994; 91:7311-7315.
91. Bukrinsky M I et al. A nuclear localization signal within HIV-1 matrix protein that governs infection of nondividing cells. Nature 1993; 365:666-669.
92. Gallay P, Swingler S, Song J et al. HIV nuclear import is governed by the phosphotyrosine-mediated binding of matrix to the core domain of integrase. Cell 1995; 83:569-576.
93. Bukrinskaya AG, Ghorpade A, Heinzinger NK et al. Phosphorylation-dependent human immunodeficiency virus type 1 infection and nuclear targeting of viral DNA. Proc Natl Acad Sci USA 1996; 93:367-371.
94. Hancock JF, Magee AI, Childs JE et al. All ras proteins are polyisoprenylated but only some are palmitoylated. Cell 1989; 57:1167-1177.
95. Hancock JF, Paterson H, Marshall CJ. A polybasic domain or palmitoylation is required in addition to the CAAX motif to localize p21ras to the plasma membrane. Cell 1990; 63:133-139.
96. Leevers SL, Paterson HF, Marshall CJ. Requirement for Ras and Raf activation is overcome by targeting Raf to the plasma membrane. Nature 1994; 369:411-414.
97. Lacal PM, Pennington CT, Lacal JC. Transforming activity of ras proteins translocating to the plasma membrane by myristylation sequence from the src gene product. Oncogene 1988; 2:533-537.
98. Lill R, Neupert W. Mechanisms of protein import across the mitochondrial outer membrane. TICB 1996; 6:56-61.
99. Verner K. Cotranslational protein import into mitochondria: an alternative view. TIBS 1993; 18:366-371.
100. Rizzuto R, Simpson AWM, Brini M et al. Rapid changes of mitochondrial Ca^{2+} revealed by specifically targeted recombinant aequorin. Nature 1992; 358:325-327.
101. Biocca S, Ruberti F, Tafani M et al. Redox state of single-chain Fv fragments targeted to the endoplasmic reticulum, cytosol and mitochondria. Biotechnology 1995; 13:1110-1115.
102. Heijne GV. Mitochondrial targeting sequences may form amphiphilic helices. EMBO J 1986; 5:1335-1342.
103. Allison DS, Schatz G. Artificial mitochondrial presequences. Proc Natl Acad Sci USA 1986; 83:9011-9015.

104. Lemire BD, Fankhauser C, Baker A et al. The mitochondrial targeting function of randomly generated peptide sequences correlate with predicted helical amphiphilicity. J Biol Chem 1989; 264:20206-20215.
105. Roise D, Schatz G. Mitochondrial presequences. J Biol Chem 1988; 263, 4509-4511.
106. Nguyen M, Millar DG, Yong VW et al. Targeting of bcl-2 to the mitochondrial outer membrane by a COOH-terminal signal anchor sequence. J Biol Chem 1993; 268:25265-25268.
107. Krajewski S, Tanaka S, Takajama S et al. Investigations of the subcellular distribution of the bcl-2 oncoprotein: residence in the nuclear envelope, endoplasmic reticulum and mitochondrial outer membranes. Cancer Res 1993; 53:4701-4714.
108. Zhu et al. Bcl-2 mutants with restricted subcellular location reveal spatially distinct pathways for apoptosis in different cell types. EMBO J 1996; 15:4130-4141.
109. McNew JA, Goodman JM. The targeting and assembly of peroxisomal proteins: some old rules do not apply. TIBS 1996; 21:54-58.
110. Subramani S. Convergence of model systems for peroxisome biogenesis. Curr Opin Cell Biol 1996; 8:513-518.
111. Hijikata M, Wen JK, Osumi T et al. Rat peroxisomal 3-ketoacyl-CoA thiolase gene. Occurrence of two closely related but differentially regulated genes. J Biol Chem 1990; 265:4600-4606.
112. Gould S J, Keller GA, Subramani S. Identification of a peroxisomal targeting signal at the carboxy terminus of firefly luciferase. J Cell Biol 1987; 105:2923-2931.
113. Gould SJ, Krisans S, Keller GA et al. Antibodies directed against the peroxisomal targeting signal of firely luciferase recognize multiple mammalian peroxisomal proteins. J Cell Biol 1990; 110:27-34.
114. Purdue PE, Lazarow PB. Targeting of human catalase to peroxisomes is dependent upon a novel COOH-terminal peroxisomal targeting sequence. J Cell Biol 1996; 134:849-862.
115. Swinkels BW, Gould SJ, Bodnar AG et al. A novel, cleavable peroxisomal targeting signal at the amino-terminus of the rat 3-ketoacyl-CoA thiolase. EMBO J 1991; 10:3255-3262.
116. Glover JR, Andrews DW, Subramani S et al. Mutagenesis of the amino targeting signal of Saccharomyces cerevisiae 3-ketoacyl-CoA thiolase reveals conserved amino acids required for import into peroxisomes in vivo. J Biol Chem 1994; 269:7558-7563.
117. McCammon MT, McNew JA, Willy PJ et al. An internal region of peroxisomal membrane protein PMP47 is essential for sorting to peroxisomes. J Cell Biol 1994; 124:915-925.
118. Chiang H, Terlecky SR, Plant CP et al. A role for a 70-kilodalton heat shock protein in lysosomal degradation of intracellular proteins. Science 1989; 382-385.
119. Dice JF. Peptide sequences that target cytosolic proteins for lysosomal proteolysis. TIBS 1990; 15:305-309.
120. Dice JF, Agarraberes M, Kirven-Brooks et al. Heat shock 70-kD proteins and lysosomal proteolisis. In: Morimoto RI, Tissieres A, Georgopoulos C, eds. The Biology of Heat Shock Proteins and Molecular Chaperones. Cold Spring Harbor, NY: Cold Spring Harbor Laboratory Press, 1994:137-151.
121. Cuervo AM, Dice F. A receptor for the selective uptake and degradation of proteins by lysosomes. Science 1996; 273:501-503.
122. Stock D, Nederlof PM, Seemuller E et al. Proteasome: from structure to function. Curr Opin Biotech 1966; 7:376-385.
123. Ciechanover A. The ubiquitin-proteasome proteolytic pathway. Cell 1994; 79:13-21.

124. Groettrup M, Soza A, Kuckelkorn U et al. Peptide antigen production by the proteasome: complexity provides efficiency. Immunol Today 1996; 17:429-435.

125. Hochstrasser M. Ubiquitin, proteasomes, and regulation of intracellular protein degradation. Curr Opin Cell Biol 1995; 7:215-223.

126. Hilt W, Wolf DH. Proteasome: destruction as a programme. TIBS 1996; 21:96-102.

127. Bachmair A, Finley D, Varshavsky A. In vivo half-life of a protein is a function of its amino-terminal residue. Science 1986; 234:179-186.

128. Varshavsky A. The N-end rule: Functions, mysteries, uses. Proc Natl Acad Sci USA 1996; 93:12142-12149.

129. Li X, Stebbins B, Hoffman L et al. The N terminus of antizyme promotes degradation of heterologous proteins. J Biol Chem 1996; 271:4441-4446.

130. Rolfe M, Beer-Romero P, Glass S et al. Reconstitution of p53-ubiquitinylation reactions from purified components: the role of human ubiquitin-conjugating enzyme UBC4 and E6-associated protein (E6AP). Proc Natl Acad Sci USA 1995; 92:3264-3268.

131. Johnson ES, Bartel B, Seufert W et al. Ubiquitin as a degradation signal. EMBO J 1992; 11:497-505.

132. Glotzer M, Murray AW, Kirschner MW. Cyclin is degraded by the ubiquitin pathway. Nature 1991; 349:132-138.

133. Klotzbucher A, Stewart E, Harrison D et al. The 'destruction box' of cyclin A allows B-type cyclins to be ubiquinated, but not efficiently destroyed. EMBO J 1996; 15:3053-3064.

134. Holloway SL, Glotzer M, King RW et al. Anaphase is initiated by proteolysis rather than by the inactivation of maturation-promoting factor. Cell 1993; 73:1393-1402.

135. Stewart E, Kobayashi H, Harrison D et al. Destruction of Xenopus cyclins A and B2, but not B1 requires binding to P34 cdc2. EMBO J 1994; 13:584-594.

136. Rogers SW, Wells R, Rechsteiner M. Amino acid sequences common to rapidly degraded protein: the PEST hypothesis. Science 1986; 234:364-368.

137. Rechsteiner M, Rogers SW. PEST sequences and regulation by proteolysis. TIBS 1996; 21:267-271.

138. Ghoda L, Phillips MA, Bass KE et al. Trypanosome ornithine decarboxylase is stable because it lacks sequences found in the carboxyl terminus of the mouse enzyme which target the latter for intracellular degradation. J Biol Chem 1990; 265:11823-11826.

139. Rechsteiner M. PEST sequences are signals for rapid intracellular proteolysis. Seminars in Cell Biol 1990; 1:433-440.

140. Salama SR, Hendricks KB, Torner J. G1 cyclin degradation: the PEST motif of yeast CLN2 is necessary, but not sufficient for rapid protein turnover. Mol Cell Biol 1994; 14:7953-7966.

141. Litman P, Barge J, Ginsburg I. Microtubules are involved in the localization of tau mRNA in primary neuronal cultures. Neuron 1994; 13:1463-1474.

142. Ainger K, Avossa D, Morgan F et al. Transport and localization of exogenous myelin basic protein mRNA microinjected into oligodendrocytes. J Cell Biol 1993; 123:431-441.

143. Curtis D, Lehmann R, Zamore PD. Translational regulation in development. Cell 1995; 81:171-178.

144. St Johnston D The intracellular localization of messenger RNAs. Cell 1995; 81:161-170.

145. Mcdonald PM, Struhl G. Cis-acting sequences responsible for anterior localization of bicoid RNA in *Drosophila* embryos. Nature 1988; 336:595-598.

146. Mowry KL, Melton DA. Vegetal messenger RNA localization directed by a 340-nt RNA sequence element in *Xenopus* oocytes. Science 1992; 255:991-994.

Intercellular Immunization

Antonino Cattaneo, Patrizia Piccioli and Francesca Ruberti

General Principles

In this chapter we shall describe how ectopic antibody expression can be used to interfere with the function of molecules that are located in (or are accessible to) the extracellular environment of a cell or a tissue. This concept was initially introduced with the nervous system in mind[1] and will be discussed primarily in the context of studies of the central nervous system, but the principles of the approach can be extended to other systems as well.

Antibodies are normally secreted by plasma cells. It is now well established that virtually all of the nonlymphoid cells tested can support the secretion of functional immunoglobulins. This is true for both animal and plant cells. However, as already demonstrated in the early studies on antibody secretion by nonlymphoid cells,[1] the efficiency of secretion by different cell types varies dramatically by orders of magnitude. In that study, it was found in particular that the efficiency of antibody secretion by cells related to the nervous system is comparable to that of lymphoid cells transfected with the same antibody genes. Thus, it was proposed[1] that the local secretion (by cells of the nervous system) of specific monoclonal antibodies, cloned from the corresponding hybridoma cell lines, could be utilized to perform functional and developmental studies in the otherwise intact nervous system of transgenic mice (the so called neuroantibody approach).[1-3] The advent of phage technology, with new ways of isolating, selecting and engineering cloned recombinant antibodies greatly enriches the potential of this experimental approach.

How to get antibodies to be produced across the blood-brain barrier, by cells of the nervous system? Different strategies for tackling the problem can be pursued, which include the production of transgenic mice, the grafting of cells engineered to secrete antibody genes and the direct infection of neural cells with viral vectors harboring the antibody genes.

Whilst the transgenic approach, as discussed below, may find applications in a research context or in the creation of experimental models for neurological pathologies, the local expression of recombinant antibodies by cells of the nervous system may provide an experimental scenario of some therapeutical potential as it would circumvent the problem that the blood-brain barrier poses to the accessibility of a circulating antibody to the CNS. Thus, a therapeutic antibody, potentially useful for some CNS disorder, would not have access to its target in the CNS,

Intracellular Antibodies: Development and Applications, edited by
Antonino Cattaneo and Silvia Biocca. © Springer-Verlag and Landes Bioscience 1997.

if delivered systemically, and strategies to achieve the delivery of the corresponding gene for local expression in the CNS, such as the grafting of cells engineered to secrete antibodies or the use of viral vectors, would be useful.

In conclusion, the basis of the neuroantibody technique is to harness the efficient secretion of antibodies by glial and neuronal cells in order to achieve the local production of antibodies in the CNS derived from hybridoma cell lines or from phage libraries. These will act as an immunological "knife".

Neuroantibodies: Neurological Lesions Created by Antibody Secreting Neural Cells

Selective lesioning techniques are at the heart of functional neuroscience research. The availability of gene transfer techniques, in particular the ability to create lines of transgenic mice, has opened new possibilities to study the physiology of the nervous system. "Loss-of-function" mutations can be created in mice by different strategies. Study of the response of the perturbed system should provide insights into the function of the complex system. Among different methods used to inhibit the function of selected genes in mammalian organisms which are presently being pursued (discussed in chapter 2), the ectopic expression of antibodies is a recent addition and as the first transgenic studies are appearing, its merits and problems can be evaluated.

The central nervous system (CNS), because of its complexity and highly organized cytoarchitecture, represents an attractive target for the use of antibodies as perturbing agents, interfering with the action of extracellularly acting molecules. In the CNS, the importance of extracellular signaling molecules is paramount, as is the geometry of the system, and the possibility of exploiting the richness of the antibody repertoire to modulate in a controlled way the function of these molecules would be very fruitful.

Before discussing the use of ectopic expression of recombinant antibodies for intercellular immunization in the CNS (neuroantibody strategy), we would like to recall some general properties of neuronal circuits:

1. Many processes in the development and function of the nervous system occur through molecules acting in the extracellular environment or topologically accessible from the extracellular environment (e.g., neurotransmitters, neuropeptides, growth factors and their corresponding receptors, adhesion molecules).

2. Although the number of the cells in the brain is very high (in fact this number is often quoted as a measure of the complexity of the nervous system), the number of functionally different cell types in any given area is rather limited. The different neuronal cell types can be distinguished not only on the basis of their morphology, but also according to the extracellular signals they transmit and they respond to.

3. Similar local circuits involving such a limited number of cells are repeated many times in a continuous way ("modular structures").

4. The diffusion of large molecules, such as antibodies, or even antibody fragments in the brain tissue is rather limited and local accessibility of antibodies to target molecules acting at crucial sites, such as the synaptic cleft, would therefore be required in order to achieve an effective inhibition.

Functional and developmental studies on the mammalian nervous system would greatly benefit from the ability to specifically interfere with the function of selected neuronal subpopulations or pathways, by perturbing the action of such extracellular or extracellularly exposed molecules, to produce functional lesions in a controlled fashion.

The first attempts to pursue this strategy have relied on the creation of transgenic mice lines carrying a transgene encoding for the desired antibody under the control of a suitable promoter.

Different choices in terms of promoters and enhancers will enable different questions to be addressed and different experimental models to be produced. The possible choices include: (i) strong promoters without a particular tissue specificity, but permissive for expression in the CNS; one example is the cytomegalovirus early region promoter (CMV promoter);[4,5] (ii) promoters and/or enhancers with a broad specificity for the CNS, and in particular, for terminally differentiated neuronal[6-9] or glial cells;[10] these include the promoters of neuronally expressed genes such as neuronal specific enolase (NSE), α-tubulin, synapsin, neurofilament proteins, Thy-1, PDGF-B chain. Expression in glial cells may be achieved by using the promoters of glial specific genes such as the myelin basic protein (MBP) or the glial fibrillar acidic protein (GFA). All these have been used to direct the transcription of transgenes in the CNS with varying degrees of efficacy and of stringency; (iii) promoters with a restricted regional specificity for CNS subregions, or cell types; expression in postmitotic neurons of the forebrain, the hippocampus in particular, has been achieved using the promoter from the α-CaMKII gene,[11] while more restricted expression patterns of transgenes such as the retina or the cerebellum[12] can be achieved; (iv) inducible promoters such as the tetracycline regulated[13] or the ecdysone regulated[14] systems; (v) the transcription regulatory sequences of the gene, whose one product is aiming to block with the antibodies, might be conceivably used to drive the transcription of the corresponding antibody so that all cells expressing a particular gene would also express an antibody directed against its gene product. It should be considered that with promoter-transgene fusions, that the precise spatiotemporal pattern of expression of the transgene will depend also on the integration site, on the presence of enhancer or silencers close to the integration site and on the absence of such regulatory elements in the gene fusion used.

As for the antibody forms utilized, the transgenic experiments performed so far have all utilized full length antibodies. This is because the secretion of whole antibody molecules is more efficient and more predictable than that of ScFv fragments. There is no reason why in the future, other antibody forms could not be considered.

Transgenic Mice Expressing Anti-NGF Antibodies

The importance of the choice of the promoter, for the neuroantibody approach, is illustrated by a study in progress in which transgenic mice expressing antibodies against the neurotrophic factor NGF (nerve growth factor) were produced.[15]

NGF, which is now known to form part of the so called neurotrophin gene family, has an important role in the developing and mature nervous system.[16] NGF regulates survival, differentiation and maintenance of specific neuronal populations, in both the peripheral and central nervous system, and was first characterized as a target-derived survival factor for developing sympathetic and sensory neurons.[16] In the PNS, NGF is produced by muscles and other cells which make up the target field which axons grow during development. Competition for limited amounts of target derived survival signals is thought to underlie the death of many central (CNS) and peripheral (PNS) neurons during development (and the survival of successful competitors!). Minute amounts of NGF control neuronal survival of specific cell populations during the critical period of development when cellular death occurs in the nervous system. This classical paradigm of neurotrophic action by NGF has been extended to the central nervous system[17] where striatal and basal forebrain cholinergic neurons have been identified as target cells for NGF. When administered to aged, cognitive impaired rats, NGF improves memory and other behavioral responses. Moreover, the action of NGF and other neurotrophins in the CNS is broader than in the PNS being related not just to neuronal survival during development, but also to the modulation of activity-dependent synaptic plasticity in the adult after development is terminated.[18] Also, a very important role of NGF and other neurotrophins as neuroprotective agents towards a variety of brain insults has been reported. Moreover, several lines of evidence show that NGF regulates the function of nociception in the postnatal life, as well as in adulthood,[19] by modulating different phenotypic parameters of sensory afferents. These results suggested that NGF may have beneficial roles in a variety of neurodegenerative diseases or neuropathological situations and have prompted investigations into its possible therapeutical use.[20] Clinical trials using NGF, and other neurotrophic factors for the treatment of different neurological disorders are under way,[21] but they mostly involve peripheral neuropathies, not only because these are more accessible to systemic administration of the factor, but because the character of the disorders is better characterized. On the contrary, the translation of animal experiments involving NGF and other neurotrophins to the treatment of human CNS diseases is still in its infancy and more evidence is needed to strengthen the case for a therapeutic use of NGF. The lack of animal models, in which the actions of endogenous levels are competed in a chronic way in the adult CNS, is probably the main reason progress has been hindered on the field; and therefore, a great need of reproducible models to approach the problem. The knock-out by homologous recombination of the genes for NGF[22] and for its receptor TrkA[23] have been recently reported, but in both cases, the mice die after the first postnatal weeks because of severe developmental defects arising from failure of NGF to act on its well-known target cells. Thus, these knock-out models do not allow to study questions related to the role of NGF in the adult nervous system.

We decided, therefore, to exploit the neuroantibody approach and to produce transgenic mice expressing a neutralizing antibody against NGF that would be amenable to study the wide spectrum of activities of NGF. At first, the aims of this work were to demonstrate that the expression of transgenic anti-NGF antibodies could successfully compete with endogenous NGF on the classical targets, such as sympathetic neurons and, second, to achieve a ubiquitous expression also in adult mice. For this reason, the early region promoter of the human cytomegalovirus was used to direct the expression of the recombinant anti-NGF.

The starting point was the monoclonal antibody α-D11,[24] a rat IgG2a antibody which binds NGF with an affinity greater than 10^{-9} M and neutralizes very efficiently the biological action of NGF both in vitro[24] and in vivo.[25,26] This neutralizing activity is exerted by a direct competition between NGF itself and the trkA receptor. αD11 recognizes[27] the NGF loop region from residues 41-49, which forms part of the interaction surface of NGF with its receptor TrkA and distinguishes NGF from the other members of the neurotrophin family. Consistently, mAb αD11 does not crossreact[27] with the other members of the neurotrophin family BDNF, NT3 and NT4. These functional and molecular characteristics of the mAb αD11 prospect this antibody as an ideal reagent for in vivo studies.

The cloned variable regions of the α-D11 antibody[3] were linked to human γ1 constant regions to facilitate the detection of the transgenic antibody against the background of the endogenous mouse immunoglobulins, and placed under the transcriptional control of the early region promoter of the human cytomegalovirus. Mice transgenic for both transgenes (heavy and light chain genes) or for each individual transgene were derived.

Double transgenic founder mice express functional anti-NGF chimeric antibodies in the serum. Moreover, analysis of their superior cervical ganglia showed a marked immunosympathectomy, thus proving that the transgenic antibodies are effective in competing with endogenous NGF, at least in one classical target, sympathetic neurons. However, these mice did not reproduce and a double transgenic line could not be obtained. On the other hand, transgenic lines expressing only the heavy or the light chain of the αD11 were intercrossed to produce, in a two tiered approach, transgenic mice which express functional anti-NGF antibodies (Fig. 6.1A). The advantage in this approach derives from the possibility of obtaining stable lines of mice in which the expression of the transgene should not produce an NGF-deprived phenotype. In this way, two families of αD11 transgenic mice were derived (CMV-αD11-A and CMV-αD11-B), both expressing functional anti-NGF antibodies in the serum. The amount of circulating anti-NGF transgenic antibodies was in the range of 5-10 ng/ml for mice of family A, and of 50-100 ng/ml for family B. The difference in antibody levels between family A and family B can be ascribed to a difference in the levels of light chain mRNA (and hence protein) expression. High levels of anti-NGF antibodies could be demonstrated in the brain of family B mice (but not of family A mice) with a widespread spatial distribution throughout the CNS. This demonstrates that the heavy and light chain are expressed in the same set of cells. Examples of the expression of the αD11 heavy chain protein in the CNS of transgenic mice, studied by immunocytochemistry, are shown in Figures 6.2 (hippocampus) and 6.3 (cerebellum). The overall picture is that of an abundant staining of many discrete areas throughout the nervous system. The staining is mainly neuronal and is distributed throughout the cell extensions, including their cellular processes and arborizations.

Interestingly, the αD11 transgenic antibody is expressed in a developmentally regulated manner, the expression of the mRNAs for both the heavy and the light αD11 chains being about 5-fold higher at postnatal day 90 than at birth. At the protein level, the αD11 antibody concentration in the brain are one or two orders of magnitude higher in the adult than at birth (for mice of Family B). This pattern of expression is very convenient for an anti-NGF transgenic model.

The transgenic antibodies compete successfully with the action of endogenous NGF. Even in Family A transgenic mice, which express lower αD11 antibody levels, the Superior Cervical Ganglion (SCG) was found to be visibly smaller than that of controls, with a greatly reduced number of neurons.

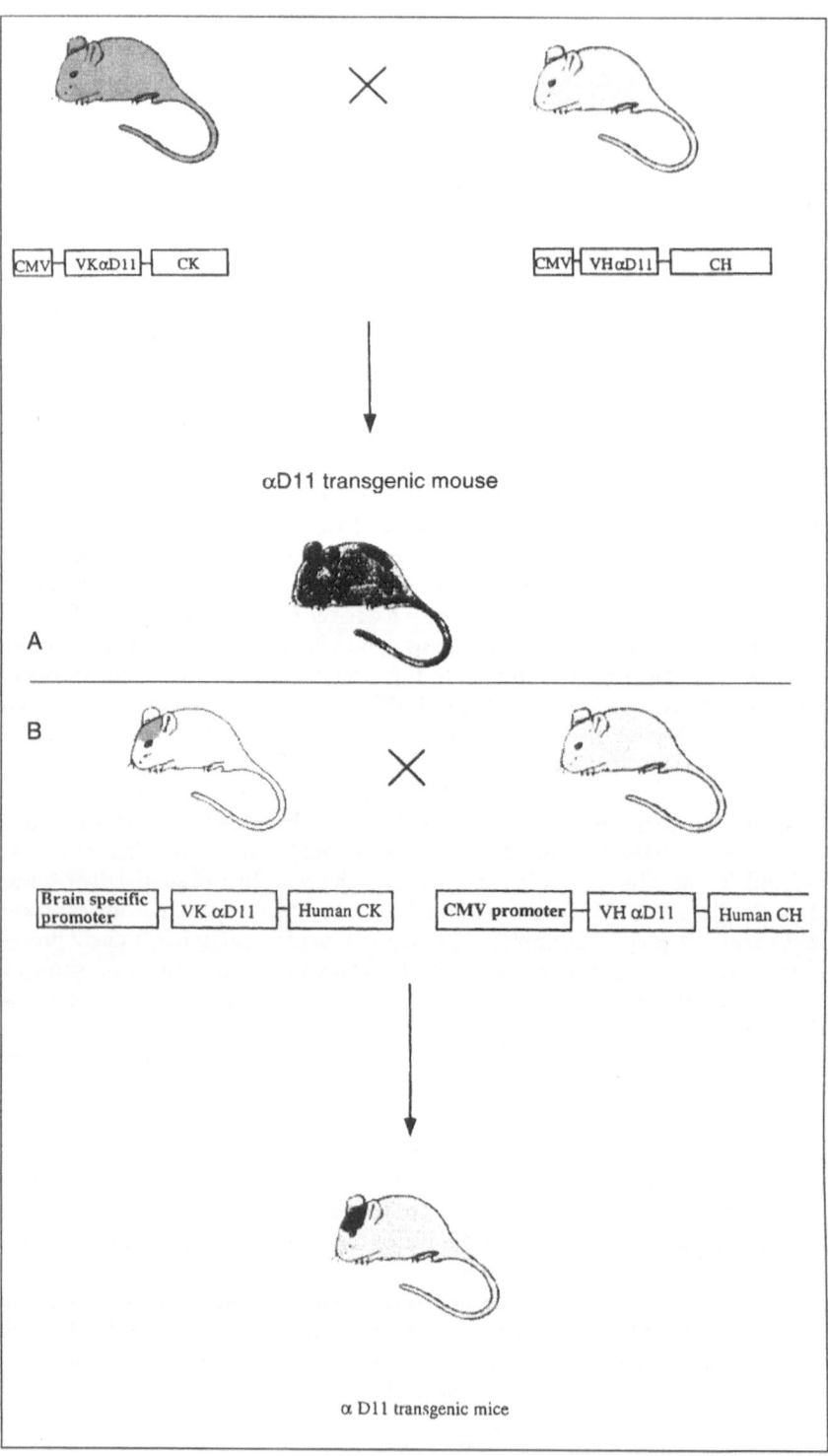

In conclusion, the production of transgenic mice which express functional anti-NGF antibody was achieved by a two-tiered approach, the intercrossing of transgenic mice which express the heavy and the light chain of the aD11 antibody. This transgenic model provides a proof of principle for the activity of transgenic antibodies, and is now amenable to study different aspects of the biological roles of NGF in the adult PNS and CNS which cannot presently be easily addressed with other transgenic models. This model represents only a first step and lends itself to further refinements. The lines of transgenic mice, expressing heavy or light α-D11 chains, could be further exploited to obtain a more refined spatial and/or temporal control of aD11 expression. This will be achieved by mating mice transgenic for one of the two chains, to new transgenic lines expressing the cognate chain under the control of a more restricted tissue specific or of an inducible promoter (Fig. 6.1B). This will provide a spatially and/or temporally more restricted expression of functional antibody.

Expression of one or both antibody chains driven by a promoter specific for expression in mammary glands (such as the promoter of the goat β-casein gene), will allow a very high expression of the recombinant antibody in the milk of lactating transgenic mice. This would allow to achieve a time controlled delivery of antibodies to neonate mice through a time controlled exposure of mice pups to transgenic lactating mothers. During this period, the blood-brain barrier is not yet fully developed in mice and the antibodies would have access to the developing nervous system.

Transgenic Mice Expressing Antibodies Against the Neuropeptide Substance P (SP)

Another application of the neuroantibody experimental strategy concerns the production and characterization of transgenic mice expressing neutralizing antibodies against the neuropeptide substance P (SP). This study provided the first demonstration of a transgenic model, in which antibodies expressed in the CNS, interfere functionally with the corresponding target molecule.[28]

SP is a peptide belonging to the tachykinin family that has been associated with the transmission of sensory and nociceptive information in the spinal cord (see 29 for a review). Antidromic release of SP from sensory afferents is responsible for vasodilatation and plasma extravasation in neurogenic inflammation. SP is also present in CNS areas such as the substantia nigra and hypothalamus, where its function is unknown and in peripheral neuronal structures, where it is involved in regulating smooth muscle motility. The actions of SP during development are

Fig. 6.1 (opposite). Intercrossing of transgenic mice expressing individual antibody chains. The NGF binding activity of the αD11 antibody requires both chains, therefore functional αD11 antibodies (black) will be assembled only if cells coexpress the light (VKαD11, dark gray) and the heavy (VHαD11, light gray) chains.

(A) The VKαD11 and the VHαD11 transgenes are expressed under the control of the CMV promoter, therefore the offspring of the intercrossing will express αD11 antibodies in all tissues.

(B) The CMV-VHαD11 mouse is crossed to a mouse expressing the VKαD11 under the control of a brain-specific promoter. The offspring will express transgenic αD11 antibodies (black) only in the brain.

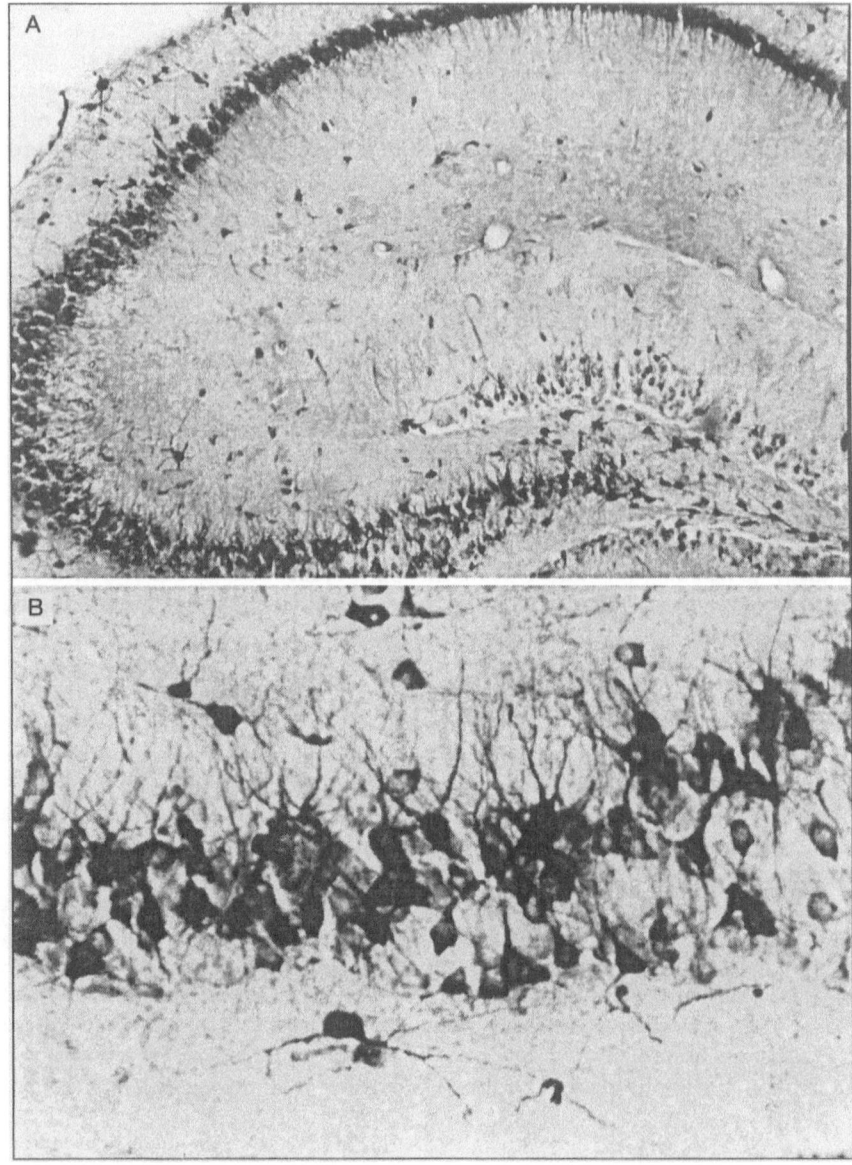

Fig. 6.2. Expression of recombinant αD11 antibody chains in the central nervous system of transgenic mice.

(A) 30 μm coronal section of brain from a 2 months old αD11 transgenic mouse, stained with antibodies against human heavy chains. Expression of the heavy chain of αD11 in the hippocampus.

(B) Higher magnification view of (A), showing neuron specific staining of hippocampal pyramidal cells.

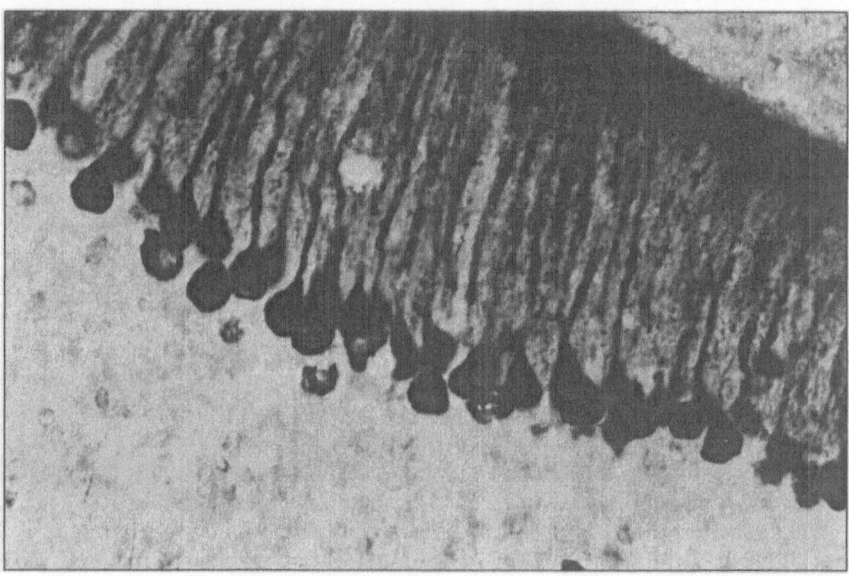

Fig. 6.3. Expression of recombinant αD11 antibody chains in the cerebellum of transgenic mice. Staining as in Fig. 6.2. Purkinje cells of the cerebellum are heavily stained.

still largely unknown. Recent data suggest a role for SP in the innervation of the developing spinal cord floor plate.[30] Roles for SP as a growth factor or as a messenger between the nervous and the immune systems have also been suggested.

The rat monoclonal antibody NC1/34HL[31] binds to SP with an nanomolar affinity, recognizes the amidated C-terminal portion of the SP peptide which is responsible for receptor binding, and does not crossreact with the related tachykinin peptides NKA and NKB. The cloned variable regions of the NC1 mAb[2] were placed under the control of the promoter of the neuroendocrine *vgf* gene[32] to engineer the expression of the chimeric NC1/34HL antibody in transgenic mice. The *vgf* gene is expressed in cells of neuronal and endocrine origin, in regions such as cerebral cortex, thalamus, hypothalamus, adenohypophysis, spinal cord and dorsal root ganglia, as well as in the adrenal medulla. Its promoter region has been characterized.[33] Many of these regions show SP and/or NK1 receptor expression as well.

Transgenic mice expressing the NC1 chimeric antibody under the transcriptional control of the *vgf* promoter are viable and have a normal lifespan. Their overall phenotype is rather mild (see below). The expression of the heavy and light NC1 antibody chains parallels that of the endogenous *vgf* gene, and peaks in the second postnatal week, after it remains at fairly high levels into adulthood. Both antibody chains are abundantly present in many discrete areas throughout the nervous system with a predominantly, if not exclusive, neuronal localization. Most important from the point of view of the neutralization of SP action(s), the antibodies are present in many CNS regions where the SP peptide is also present.

Functional antibodies will be found in vivo only if the two antibody chains are coexpressed in the same cells. The presence of transgenic SP-binding antibodies was demonstrated in the serum and the brain of transgenic mice, demonstrating assembly (and therefore coexpression) of the two antibody chains. The adrenal medulla is the probable source for the transgenic antibodies found in serum. The levels of SP binding transgenic antibodies, in serum and brain, increase after birth up to a maximum value around the second to third postnatal week, with an average value of 60 ng/ml in serum and of 700 ng/100mg in brain. These values are comparable to those obtained for the anti-NGF mice.

The anti-SP antibodies were found to be widely expressed in the CNS of transgenic mice. In order to verify from a functional point of view whether the transgenic antibodies affect some of the different systems in which the SP peptide is involved, we tested in these mice acute nociceptive behavior (tail immersion and hot plate test), neurogenic inflammation upon mustard application on the skin and motor activity and exploratory behavior in the open field test (Fig. 6.4). For practical reasons, the behavioral tests were performed on adult mice even if the levels of the antibody are lower in adult than in younger animals.

The reaction time in the hot plate test and the tail withdrawal latencies did not differ in anti-SP transgenic mice with respect to age matched control mice (Fig. 6.4A). On the contrary, mustard-oil induced plasma protein extravasation was significantly reduced in mice expressing the substance P antibody (Fig. 6.4B). Tests for motor activity were also performed showing a marked inhibition of locomotory activity, for the transgenic mice in the open field test, with a very significant reduction of line crossings and rearings (Fig. 6.4C).

In conclusion, the expression of recombinant antibodies directed against the neuropeptide substance P, under the transcriptional control of the promoter of the neuronal *vgf* gene, is functionally effective in neutralizing the actions of the neuropeptide.

The neuropeptide SP has multiple actions in the peripheral and central nervous system, some of which have been unequivocally demonstrated, while others are more controversial or unknown. For instance, the involvement of substance P as the primary neurotransmitter in acute nociception has been recently questioned, and the role of the peptides is more likely linked to chronic pain.[34] The lack of inhibition in the acute nociceptive behavioral test is in line with the lack of effect of recently introduced CNS active nonpeptide NK1 antagonists in this phenomenon.[35] On the other hand, the primary role of SP in neurogenic inflammation, smooth muscle contraction and in vasodilation is well established.[36] The observed inhibition of neurogenic inflammation in the transgenic mice confirms the validity of the transgenic model. As for the central nervous system, where the SP peptide and its receptor are widely distributed, their functional role(s) are far from clear. It has been previously shown that central injections of substance P induce locomotion activity.[37,38] Consistently, transgenic mice expressing anti substance P antibodies show a decreased locomotion activity and also a decreased exploration behavior. The precise site of action of the antibodies in causing these motor deficits remains to be determined. Answers to this question will be provided by the study of lines of transgenic mice with a different transcriptional control of the transgenic antibodies.

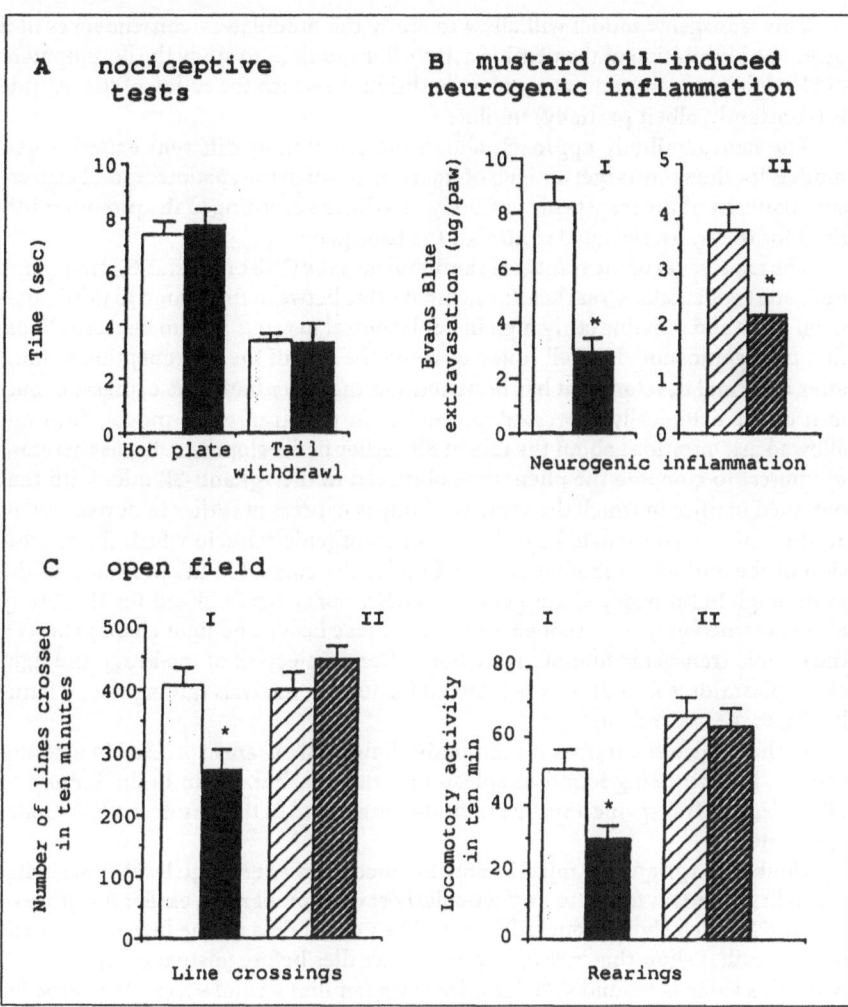

Fig. 6.4. Functional tests of SP action in vgf anti-SP transgenic mice. Thermonociception (A) was evaluated by the reaction time in the hot-plate test and by tail withdrawal latency from 50°C water. Neurogenic plasma protein extravasation (B) was evoked in the hindpaw by challenging with 5% mustard oil in paraffin oil (part I, vgf anti-SP mice, part II, Ig anti-SP mice, family #8). Motor activity (C) was evaluated in the open field test by counting line crossings and rearings during a 10 min observation period. Data represent means ± SEM (asterisk, p < .05); N = 8 for each group). In each pair of histograms, at left are the controls, at right the transgenic mice. Reprinted with permission from ref. 28.

This transgenic model will allow to study the modulatory consequences of a prolonged inhibition of the peptide action. For example, to study the development of SP-ergic synaptic pathways under conditions in which the action of the peptide is (constantly, albeit partially) inhibited.

The neuroantibody approach allows the creation of different experimental models for the same target antigen of interest, in which the spatiotemporal expression patterns of the transgenic antibody can differ according to the promoter utilized for the transcriptional control of the transgene.

The expression of the transgenic antibodies in the CNS of *vgf* anti-SP transgenic mice starts postnatally, reaches its maximal value between the second to third postnatal week and remains fairly high in adult central nervous system neurons. Thus, that transgenic model is well suited to study the role of the neuropeptide in mice after a normal development has occurred. On the other hand, the expression pattern of the ectopically expressed antibodies in that transgenic model, does not allow to ask questions about the role of SP earlier in development. It was therefore of interest to compare the phenotype observed in the *vgf* anti-SP mice with that obtained in mice in which the same antibody is expressed earlier in development. To this aim, we constructed another set of transgenic mice, in which the expression of the anti-SP antibodies is placed under the control of the promoter of the immunoglobulin heavy chain gene. The DNA constructs utilized for this study also contained enhancer sequences of the mouse heavy and light chain genes. Of the double transgenic founder mice born after coinjection of the heavy and light chain plasmids, 5 founders were selected for further analysis (giving rise to families #1, #2, #6, #8 and #9).

In these mice, the transgenic antibody chain mRNAs are expressed in a tissue-specific fashion being found in spleen and thymus, but not in brain, kidney or other tissues. Transgenic anti-SP antibodies are found in the serum at all postnatal ages tested.

Unlike the *vgf*-anti-SP mice, Ig-anti-SP mice show a very high level of perinatal mortality. This was found to be particularly severe for all mice, except for those of family #8, which show a longer lifespan. The pooled data on the lifespan for each of the families show that 75% of family #1 mice dies before postnatal day 10 (P10), while this value is around 50% for mice from families #2 and #9, (and only 5% for mice from family #8). Thus, a very small percentage of Ig-anti-SP transgenic mice live for more than 2 months. This is in sharp contrast with the mild phenotype observed with the *vgf*-anti-SP mice. Transgenic mice which do survive after the first postnatal weeks are much smaller than age-matched controls.

The severity of the phenotype observed, points to an important role for the neuropeptide SP during mouse development. The important question arises as to why the Ig-anti-SP mice show such a dramatic lethal phenotype, while the *vgf*-anti-SP do not, even if in the latter, the transgenic antibodies are effective in blocking some of the actions of the SP neuropeptide in adults.[28] One main difference between the two lines of mice lies in the tissue distribution of the transgenic antibodies, with a prevalent neuronal expression in the *vgf*-anti-SP mice, and a prevalent expression in lymphoid organs, and hence in the serum, in the Ig-anti-SP mice. However, the levels of transgenic anti-SP antibodies in the serum of *vgf*-anti-SP mice are very similar to those of the Ig-anti-SP mice at all the postnatal ages tested. Thus, it is unlikely that the different phenotypes observed in the two transgenic lines are due to differences in the postnatal expression levels of the circulating antibodies. On the other hand, the developmental onset of expression of the transgenic antibodies is different in the two lines of mice. Very high levels of anti-SP antibodies are found in embryos (embryonic age E16-E19) of all Ig-anti-SP mice (except for those of family #8), while no significant amount of anti-SP chimeric antibody can be detected in vgf-anti-SP embryos, nor in family #8 Ig-anti-SP embryos. Thus, there appears to be a direct correlation between the presence of high levels of transgenic antibodies between embryonic days E16 and E19 and the subsequent fate of the litters: high perinatal lethality for family #9 Ig-anti-SP mice, less than 5% perinatal lethality for family #8 Ig-anti-SP and for *vgf*-anti-SP mice.

In adult *vgf*-anti-SP mice, the expression of the transgenic antibodies induces a marked inhibition of neurogenic inflammation and of general motor behavior.[28] In mice from the family #8 Ig-anti-SP, which do not show a lethal phenotype, neurogenic inflammation is also inhibited (Fig. 6.4) showing that the antibodies are effective in counteracting the actions of the neuropeptide in adult mice from this family as well. On the other hand, motor activity is not influenced, suggesting that the inhibition observed for the *vgf* anti-SP mice is due to antibodies acting in the CNS.

These results point to a fundamental, yet still unknown, role of the neuropeptide SP during embryonic development. In principle, the lethal phenotype observed in the Ig-anti-SP mice may be caused by the antibody-mediated disruption of the interactions of SP with the NK-1 receptor (NK1-R) or with alternative receptors. While most of the well characterized actions of SP are mediated by the NK-1 R receptors, non-NK1-R mediated actions have been described in neuronal[39-41] and non-neuronal[42] cells. For instance, SP is known to inhibit nicotinic acetylcholine receptors, by interacting directly in a noncompetitive way with the β subunit of the neuronal nAChR.[41] Also, some non-neuronal actions of SP, such as those on monocytes, are not mediated by the NK1 receptors (NK1R).[42] Interestingly, transgenic mice in which the gene for the substance P receptor NK1-R has been disrupted by gene targeting have recently been derived.[43] These mice show a normal development, are healthy and do not show the lethal phenotype displayed by the Ig-anti-SP transgenic mice. Therefore, it may be inferred that the lethal phenotype in these mice is not due to block of the SP-NK1R interaction, but rather to the non-NK1R-mediated actions of the neuropeptide which may turn out to be physiologically very important.

To summarize, the results obtained with the transgenic mice expressing antibodies against the neuropeptide SP, as well as antibodies against NGF, provide proofs of principle of the validity of the neuroantibody approach. Changing the spatiotemporal expression of a recombinant antibody, by the use of different promoters, can lead to very different transgenic models. These may lend themselves to study different physiological aspects of one given target molecule.

Refining the Secretion of Antibodies

In the transgenic studies performed so far, antibodies are constitutively secreted from neuronal cells and act by competing with the extracellular pool of their target antigens. An intracellular interaction between antigen and antibody is in principle possible, if the two are coexpressed. This would make the inhibition more stringent, not relying only on extracellular interactions, but also on intracellular ones. For NGF and SP, an intracellular interaction with the antibodies is unlikely since both NGF and SP are processed into their mature form from larger precursors pre-pro-proteins. This processing occurs in specialized compartments of the secretory pathway, therefore, the unprocessed antigen is not necessarily recognized by the antibody. Moreover, both NGF and SP, are secreted by neurons through the regulated secretory pathway. Secretory antibodies could be re-routed from the constitutive to the regulated secretory pathway, by appropriate targeting signals such as that of pro-opiomelanocortin.[44] The signal for targeting to the regulated pathway appears to be dominant over that for the constitutive secretion. This was shown[45] by injecting in PC12 cells the mRNA from a hybridoma, producing an antibody against secretogranin I. The secretogranin I antibody was packaged into secretory granules and was released by regulated exocytosis.[45] Thus, a constitutive secretory antibody can be diverted to the regulated pathway of secretion by its protein-protein interaction with a regulated secretory protein. This could be exploited further.

Engineering the regulated secretion of antibodies in combination with expression controlled by inducible promoters, should allow tight control of antibody secretion if needed. For instance, this could apply to the regulated secretion of immunotoxins (see below) in response to external stimuli of various nature.

Competition with a secreted antigen can therefore be approached by a combination of an "extracellular" mode of action with an "intracellular" one (intracellular anchors), to block the antigen from within the cell.

If the strategy of blocking the secretion of the target molecule is pursued, this should exploit anchoring or retention sequences as discussed in chapters 5 and 7. This approach has been used to block the appearance of membrane proteins to the plasma membrane,[46-48] but has not yet been applied to blocking the secretion of a soluble protein. We have shown that SEKDEL-tagged antibodies can successfully prevent the secretion of the 7B2 chaperonin (A. Braks and AC, unpublished). In the intracellular anchor mode of action, the antibody does not have to be a neutralizing one. The only binding prerequisite is that the target antigen is in a form recognizable by the antibody in the ER.

The morphological complexity of neuronal cells and, in particular, the spatial separation between functional compartments of these cells, opens a further challenge for the cell biologist interested in carefully controlling the secretion of antibodies from neurons: is it possible to achieve a localized secretion of antibodies by a neuronal cell, for instance at sites close to the synaptic cleft? In principle, secretion into the synaptic cleft could be achieved presynaptically or postsynaptically. We could envisage to implement the first possibility by engineering the targeting of antibodies to synaptic vesicles, possibly by constructing antibody fusions with proteins of the synaptic vesicles. The secretion from the postsynaptic end of a synapse (dendrites) could exploit the targeting of some mRNAs to dendrites and the protein synthesis machinery present in dendrites. Incorporating mRNA targeting sequences in the antibody expression vector (see chapter 5) to direct the mRNA to dendrites, could provide a further level of localization control and help obtaining a localized secretion of antibodies by neurons.

Grafting of Engineered Cells

Besides the production of transgenic models, another approach to achieve the local production of antibodies across the blood-brain barrier by cells of the nervous system, involves the transplantation of genetically engineered cells into the CNS. This approach may have some therapeutic potential.

Brain tissue grafts have been employed for many years as a tool to study aspects of developmental and functional neurobiology, exploiting the fact that the brain is in some respects, "an immunological privileged site". Grafted cells can exert a functional influence over the host brain by various mechanisms including: (i) the release of growth and trophic factors of deficient neurochemicals or hormones (the paracrine model) and (ii) the precise reinnervation and reformation of synaptic connections between the neurons of the graft and host brain.

More recently, grafted cells have been engineered to express and produce a molecule of interest, such as a neurotrophic factor.[49] Accordingly, grafted cells could be engineered to locally secrete a recombinant antibody.

This ex vivo approach is not different in principle, to ex vivo approaches that are being pursued for various human diseases, including tumor and viral disease therapy: the appropriate (autologous or heterologous) cells are transduced in vitro to express the therapeutic gene of interest, and transduced cells are reintroduced in vivo. The choice of the engineered donor cell to be grafted back into the CNS is an important issue. In rat and primate models, genetically modified autologous fibroblasts or genetically engineered xenogenic cells isolated within a permeability-selective polymer capsule that restricts cell growth, have been successfully used to deliver exogenous gene products.

Advances have been made recently in the biology of stem cells in relation to the nervous system. Stem cells from the CNS of mouse, rat or human can be isolated from various regions of the developing CNS. These cells divide in response to epidermal growth factor or other mitogens,[50-52] and following removal of the mitogen and exposure to a suitable substrate, differentiate into neurones, astrocytes and oligodendrocytes. Also totipotent mouse embryonic stem cells (ES) can be expanded in culture in the presence of the appropriate factors and made to differentiate into neural tissue. Both types of stem cells have been transplanted into the adult brain and in principle, could represent suitable hosts for the secretion of gene products of interest, including antibodies. Perhaps one of the greatest challenges at present is to fully characterize the various types of CNS stem cells driven by different mitogens in vitro with regard to their proliferative capacity, differentiation potential and possibility of being engineered by (retro)viral infection or transfection prior to grafting. Once these questions will receive adequate answers, CNS stem cells could provide the preferred source of neural tissue for transplantation.

Direct delivery to the CNS, with viral vectors, is another possible strategy to achieve local secretion of antibodies in a gene therapy perspective. This is a problem in common with the delivery of other therapeutic genes potentially useful for CNS disorders.[53] It is well recognized that the CNS presents unique challenges as an environment for therapeutic gene delivery, partly, but not only, because of the blood-brain barrier. The ideal vector system is not yet available. Retroviruses are ineffective for introducing genes in the mitotically quiescent cells of the CNS. A retroviral vector system based on the human immunodeficiency virus has recently been described,[54] capable of stable gene transfer to CNS neurons in vivo in adult rats, but its potential needs to be further evaluated. Gene delivery to peripheral and central neurons has been achieved with herpes simplex virus-type 1[55] and adenovirus vectors.[56] Recombinant adeno-associated virus (rAAV) vectors are presently being actively considered[57] and may have some advantages over other systems currently used to deliver genes to the CNS. Among these, AAV vectors are capable of transducing and eliciting long-term stable expression from postmitotic cells and produce little inflammatory response. Any of these vectors could be used to express antibody genes in the CNS with the same indications as for other genes.

In conclusion, it is clear that the local expression of antibodies within the CNS, obtained by the grafting of engineered cells or by direct delivery of the corresponding genes would allow to circumvent the obstacle posed by the blood-brain barrier to circulating antibodies. This may increase the range of targets for potentially therapeutic antibodies.

Intercellular Immunization for Cell Ablation

Antibodies engineered for schemes of intercellular immunization can be equipped with effector functions (see chapter 7) that, combined with their binding properties, transform them into more powerful reagents. In particular, we shall describe here effector functions that, following the binding event mediated by the antibody, lead to the death of the cell recognized by the antibody. The context in which these studies have been developed is that of cancer therapy, but many of the concepts and techniques can in principle, be applied to other systems. Thus, the specificity provided by the antibody variable regions can be harnessed and used to direct (i) cytotoxic T lymphocytes or (ii) fusion toxins towards tumor cells.

In the first approach,[58-62] the variable regions of an antibody have been combined to the constant regions of T cell receptor chains, resulting in chimeric genes endowing T lymphocytes expressing them with antibody-type recognition (i.e., non-MHC restricted). In this case, the effector function is provided by the activation of the T cell following the binding event. For instance, antigen specific ScFv fragments were linked to γ or ζ chain, the signal-transducing subunits of T cell receptor and the chimeric proteins were expressed on the membrane receptor on a cytolytic T cell. Upon antigen binding, these cells produced interleukin 2 and were able to lyse target cells in an antigen-dependent way. Such chimeric antibody fusions can be exploited to provide T cells and other effector lymphocytes with antibody-type recognition directly coupled to cellular activation. In principle, this type of scheme could be extended to the antigen-antibody mediated activation of different cell types, providing effector functions that may not be limited to target cell lysis.

Target cell lysis is, however, a particularly powerful and attractive effector function. Antibody-mediated cell lysis has also been achieved by engineering the so called "immunotoxins".[63,64] These are fusion proteins in which the binding moiety of natural toxins is substituted by antibodies which are fused or coupled to toxic subunits. In this volume, we did not deal with immunotoxins since these are normally delivered as proteins, produced in *E. coli* or by chemical cross-linking. Two recent papers[65,66] describe how mammalian cells can be genetically modified to produce and secrete targeted toxin proteins while remaining viable. This was done by fusing a ScFv fragment or a light chain (to be expressed with the corresponding heavy chain) to truncated Pseudomonas exotoxin A encompassing the translocation and the toxic domains. By demonstrating that recombinant immunotoxins can be secreted by mammalian cells, these authors show that it is feasible to produce, in vivo, targeted toxins to destroy specific cells. In one application of this strategy,[65] HIV-specific killer cells were generated by genetically modifying lymphocytes and shown to have potent and selective cytotoxicity to the targeted HIV-infected cells in vitro. In another application,[66] tumoral cells were selectively killed by cocultivating them with cells expressing the toxin fused to an antibody directed against a membrane antigen of tumor cells.

Using antibodies secreted from one cell type to deliver a toxic gene to another cell population in an intercellular mode of communication, is a very powerful application of intercellular immunization. The production of the recombinant immunotoxin may be potentially controlled at different levels, including regulated transcription and regulated secretion. The toxicity itself may be controlled the delivered gene is the herpes virus thymidine kinase gene. Cells receiving and expressing this gene will then become susceptible to the killing action of gancyclovir.

In general, the possibility of expressing immunotoxins in mammalian cells adds a new dimension to the possible schemes of intercellular immunization: antibody mediated cell to cell delivery of toxic or beneficial genes will find many applications.

References

1. Cattaneo A, Neuberger MS. Polymeric immunoglobulin M is secreted by transfectants of nonlymphoid cells in the absence of immunoglobulin J chain. EMBO J 1987; 6:2753-2758.
2. Piccioli P, Ruberti F, Biocca S et al. Neuroantibodies: molecular cloning of a monoclonal antibody against substance P for expression in the central nervous system. Proc Nat Acad Sci USA 1991; 88:5611-5615.
3. Ruberti F, Bradbury A, Cattaneo A. Cloning and expression of an anti-NGF antibody for studies using the neuroantibody approach. Cell Mol Neurobiol 1993; 13:559-568.
4. Boshart MF, Weber G, Jahn K et al. A very strong enhancer is located upstream of an immediate early gene of human cytomegalovirus. Cell 1985; 41:521-530.
5. Baskar JF, Smith PP, Nilaver G. The enhancer domain of the human cytomegalovirus major immediate early promoter determines cell type specific expression in transgenic mice. J Virol 1996; 70:3207-3214.
6. Forss-Petter S, Danielson PE, Catsicas S et al. Transgenic mice expression of β-galactosidase in mature neurons under neuron-specific enolase promoter control. Neuron 1990; 5:187-197.
7. Vidal M, Morris R, Grosveld F et al. Tissue specific control elements of the Thy 1 gene. EMBO J 1990; 9:833-840.
8. Lewis SA, Cowan NJ. Anomalous placement of introns in a member of the intermediate filament multigene family: an evolutionary conundrum. Mol Cell Biol 1986; 6:1529-1534.
9. Sasahara M et al. PDGF B-chain in neurons of the central nervous system, posterior pituitary and in a transgenic model. Cell 1991; 64:217-227.
10. Readhead C, Popko B, Takahashi N et al. Expression of a myelin basic protein gene in transgenic sheeverer mice: correction of the dysmyelinating phenotype. Cell 1987; 48:703-712.
11. Mayford M, Wang J, Kandel E et al. CaMKII regulates the frequency-response function of hippocampal synapses for the production of both LTD and LTP. Cell 1995; 81, 891-904.
12. Oberdick J, Smeyne RJ, Mann JR et al. A promoter that drives transgene expression in cerebellar Purkinje and retinal bipolar neurons. Science 1990; 248:223-226.
13. Furth PA, St Onge L, Boger H et al. Temporal control of gene expression in transgenic mice by a tetracycline-responsive promoter. Proc Nat Acad Sci USA 1994; 91:9302-9306.
14. No D, Yao TP, Evans RM. Ecdysone-inducible gene expression in mammalian cells and transgenic mice. Proc Nat Acad Sci USA 1996; 93:3346-3351.

15. Ruberti F. Use of anti-NGF monoclonal antibodies to study synaptic plasticity in the developing and adult hippocampus. PhD thesis, SISSA 1966.

16. Levi-Montalcini R. The nerve growth factor 35 years later. Science 1987; 237:1154-1162.

17. Thoenen H. The changing scene of neurotrophic factors. TINS 1991; 14:165-170.

18. Thoenen H. Neurotrophins and neuronal plasticity. Science 1995; 270:593-598.

19. Lewin GR, Mendell LM. Nerve growth factor and nociception. TINS 1993; 9:353-359.

20. Lindsay RM, Wiegand SJ, Altar CA et al. Neurotrophic factors: from molecule to man. TINS 1994; 17:182-190.

21. Schaetzl HM. Neurotrophic factors: ready to go? TINS 1995; 18:463-464.

22. Crowley C, Spencer SD, Nishimura MC et al. Mice lacking nerve growth factor display perinatal loss of sensory and sympathetic neurons yet develop basal forebrain cholinergic neurons. Cell 1994; 76:1001-1011.

23. Smeyne RJ, Klein R, Schnapp A et al. Severe sensory and sympathetic neuropathies in mice carrying a disrupted Trk/NGF receptor gene. Nature 1994; 368:246-251.

24. Cattaneo A, Rapposelli B, Calissano P. Three distinct types of monoclonal antibodies after long term immunization of rats with mouse NGF. J Neurochem 1988; 50:1003-1010.

25. Berardi N, Cellerino A, Domenici L et al. Monoclonal antibodies to nerve growth factor affect the postnatal development of the visual system. Proc Nat Acad Sci USA 1994; 91:684-688.

26. Molnar M, Ruberti F, Domenici L et al. A critical period in the sensitivity of basal forebrain cholinergic neurons to NGF deprivation. Neuroreport 1997; (in press).

27. Gonfloni S. Recombinant antibodies as structural probes for neurotrophins. PhD thesis, SISSA 1995.

28. Piccioli P, Di Luzio A, Amann R et al. Neuroantibodies: ectopic expression of a recombinant anti-substance P antibody in the central nervous system of transgenic mice. Neuron 1995; 15:373-384.

29. Otsuka M, Yoshioka K. Neurotransmitter functions of mammalian tachykinins. Physiol Rev 1993; 73:229-307.

30. De Felipe C, Pinnock RD, Hunt SP. Modulation of chemotropism in the developing spinal cord by substance P. Science 1995; 267:899-902.

31. Cuello C, Galfre G, Milstein C. Detection of substance P in the central nervous system by a monoclonal antibody. Proc Nat Acad Sci USA 1979; 76:3532-3536.

32. Levi A, Eldridge J, Paterson BM. Molecular cloning of a gene sequence regulated by nerve growth factor. Science 1985; 229:393-395.

33. Possenti R, Di Rocco G, Nasi S et al. Regulatory elements in the promoter region of vgf, a nerve growth factor-inducible gene. Proc Nat Acad Sci USA 1992; 89:3815-3819.

34. Dray A, Urban L, Dickenson A. The pharmacology of chronic pain. TiPS 1994; 15:190-197.

35. Garces YI, Rabito SF, Minshall RD et al. Lack of potent antinociceptive activity by substance P antagonist CP-96,345 in the rat spinal cord. Life Sci 1993; 52:353-360.

36. Lembeck F, Donnerer J, Tsuchiya M et al. The nonpeptide tachykinin antagonist CP-96,345, is a potent inhibitor of neurogenic inflammation. Br J Pharmacol 1992; 105:527-530.

37. Kelley AE, Iversen SD. Substance P infusion into the substantia nigra of the rat: behavioral analysis and involvment of striatal dopamine. Eur J Pharmacol 1979; 60:171-179.

38. Naranjo JR, Del Rio J. Locomotor activity induced in rodents by substance P and analogues. Neuropharmacol 1984; 23:1167-1171.

39. Clapham DE, Neher E. Substance P reduces acetylcholine-induced currents in isolated bovine chromaffin cells. J Physiol 1984; 347:255-277.

40. Simasko SM, Durkin JA, Wieland GA. Effect of substance P on nicotinic acetylcholine receptor function in PC12 cells. J Neurochem 1987; 49:253-260.

41. Stafford GA, Oswald RE, Wieland GA. The β subunit of neuronal nicotinic acetylcholine receptors is a determinant of the affinity for substance P inhibition. Mol Pharmacol 1994; 45:758-762.

42. Kavelaars A, Broeke D, Jeurissen F et al. Activation of human monocytes via a non-neurokinin substance P receptor that is coupled to Gi protein, calcium, phospholipase D, MAP kinase and Il-6 production. J Immunol 1994; 153:3691-3699.

43. Bozic CR, Lu B, Hoepken UE et al. Neurogenic amplification of immune complex inflammation. Science 1996; 273:1722-1725.

44. Cool DR, Fenger M, Snell CR et al. Identification of the sorting signal motif within pro-opiomelanocortin for the regulated secretory pathway. J Biol Chem 1995; 270:8723-8729.

45. Rosa P, Weiss U, Pepperkok R et al. An antibody against secretogranin I (chromogranin B) is packaged into secretory granules. J Cell Biol 1989; 109:17-34.

46. Marasco WA, Haseltine WA, Chen SY. Design, intracellular expression, and activity of a human anti-human immunodeficiency virus type 1 gp120 single-chain antibody. Proc Natl Acad Sci USA 1993; 90:7889-7893.

47. Beerli RR, Wels W, and Hynes NE. Autocrine inhibition of the epidermal growth factor receptor by intracellular expression of a single-chain antibody. Biochem Biophys Res Comm 1994; 204:666-672.

48. Richardson JH, Sodroski JG, Waldmann TA et al. Phenotypic knockout of the high-affinity human interleukin 2 receptor by intracellular single-chain antibodies against the α subunit of the receptor. Proc Natl Acad Sci USA 1995; 92:3137-3141.

49. Gage FH, Ray J, Fisher LJ. Isolation, characterization and use of steam cells from CNS. Annu Rev Neurosci 1995; 18:159-192.

50. Cattaneo E, McKay R. Identifying and manipulating neurinal stem cells. TINS 1991; 14:338-340.

51. Weiss S, Reynolds BA, Vescovi AL et al. Is there a neural stem cell in the mammalian forebrain? TINS 1996; 19:387-393.

52. Svendsen CN, Rosser AE. Neurones from stem cells? TINS 1995; 18:465-467.

53. Karpati G, Lochmuller H, Nalbantoglu J et al. The principles of gene therapy for the nervous system. TINS 1996; 19:49-54.

54. Naldini L et al. In vivo gene delivery and stable transduction of nondividing cells by a lentiviral vector. Science 1996; 272:263-267.

55. Federoff HJ, Geschwind MD, Geller AI et al. Expression of NGF in vivo from a defective HSV-1 vector, prevents effects of axotomy on sympathetic neurons. Proc Nat Acad Sci USA 1992; 89:1636-1640.

56. Akli S et al. Transfer of a foreign gene into the brain using adenovirus vectors. Nat Genet 1993; 3:224-228.

57. Peel AL, Zolotukhin S, Schrimsher GW et al. Efficient transduction of green fluorescent protein in spinal cord neurons using adeno-associated virus vectors containing cell type specific promoters. Gene Therapy 1997; 4:16-24.

58. Gross G, Waks T, Eshhar Z. Expression of immunoglobulin-T-cell receptor chimeric molecules as functional receptors with antibody type specificity. Proc Nat Acad Sci USA 1989; 86:10024-10028.

59. Eshhar Z, Waks T, Gross G et al. Specific activation and targeting of cytotoxic lymphocytes through chimeric single chains consisting of antibody-binding domains and the gamma or zeta subunits of the immunoglobulin and T-cell receptors. Proc Nat Acad Sci USA 1993; 90:720-724.

60. Lustgarten J, Eshhar Z. Specific elimination of IgE production using T cell lines expressing chimeric T cell receptor genes. Eur J Immunol 1995; 25:2985-2991.

61. Hwu P, Yang JC, Cowherd R et al. In vivo antitumor activity of T cells redirected with chimeric antibody/T-cell receptor genes. Cancer Res 1995; 55:3369-3373.
62. Stancovski I, Schindler DG, Waks T et al. Targeting of T lymphocytes to Neu/HER2-expressing cells using chimeric single chain Fv receptors. J Immunol 1993; 151:6577-6582.
63. Pastan I, FitzGerald D. Recombinant toxins for cancer treatment. Science 1992; 254:1173-1177.
64. Wiley RG. Neural lesioning with ribosome-inactivating proteins: suicide transport and immunolesioning. TINS 1992; 15:285-290.
65. Yang AG, Chen SY. A new class of antigen-specific killer cells. Nature Biotechnology 1997; 15:46-51.
66. Chen SY, Yang AG, Chen JD et al. Potent antitumour activity of a new class of tumour-specific killer cells. Nature 1997; 385:78-81.

CHAPTER 7

Intracellular Immunization

Silvia Biocca and Antonino Cattaneo

In the previous chapters, evidence has been provided that whole antibody molecules can be efficiently assembled and secreted by a wide variety of nonlymphoid cells of animal and plant origin. Secretion of recombinant antibodies can therefore be used for schemes of intercellular immunization i.e. to compete with the action of extracellular antigens (chapter 6). If on the contrary, the antigen to be neutralized is intracellular, antibody chains or domains need to be retargeted and it becomes extremely important to know whether the retargeted antibody chains fold and assemble correctly in the different intracellular compartments. Indeed, the answer to this question represents an absolute requirement for the whole approach.

This chapter will provide answers to the following questions: (i) can an individual antibody chain be correctly targeted to the desired intracellular compartment? (ii) do the retargeted antibody chains fold correctly, so as to assemble with the cognate chain (in the case of whole antibodies) and to preserve the antigen binding specificity? (iii) do the intracellular antibodies interact in vivo with the target antigen and neutralize its action in the different subcellular compartments?

Targeting of Intracellular Antibodies

The property of dominant and autonomous targeting sequences to confer a new localization to a reporter protein has been exploited to redirect individual antibody chains to different intracellular compartments. A number of different sorting signals could in principle be used for antibody targeting as described in chapter 5. Table 7.1 reports some of the signals that have been successfully used so far for targeting antibodies or antibody domains to different intracellular compartments.[1]

Secretory Leaders

Antibodies, as secreted proteins, possess the leader sequence for secretion. Therefore, no addition of a targeting signal is necessary to obtain their secretion in any cell type so far tested. Although there appears to be no general consensus sequences for secretion, the leaders of secretory proteins are generally composed of conserved hydrophobic amino acids. However, different leader sequences may prove more or less efficient with a certain degree of species specificity in a broad sense. Immunoglobulin leader sequences have so far proven efficient to target immunoglobulin chains or domains to the secretory pathway of all nonlymphoid mammalian cells tested, while in other species such as *E. coli*, yeast or plants leader

Intracellular Antibodies: Development and Applications, edited by Antonino Cattaneo and Silvia Biocca. © Springer-Verlag and Landes Bioscience 1997.

Table 7.1. Peptide signals used for antibody targeting

Compartment	Signal	References
secretory	hydrophobic leader sequence at N-terminal: MGWSCIILFLVATATGVHSQ (for example)	many
cytoplasmic	hydrophilic leader sequence at N-terminal: MGWSKRRSSEETATAGVHSQ or leader-less	11,12,14,15, 16,17,18,21
nuclear	nuclear localization sequence at N-terminal or C-terminal: MGWSCPKKKRKVGGGTATVHSQ[a]	11,17,21
mitochondrial	presequence of mitochondrial proteins at N-terminal: MSVLTPLLLRGLTGSARRLPVPRAK[b]	3
endoplasmic reticulum (ER)	wild type leader sequence and SEKDEL[c] sequence at C-terminal	3,4,5
ER lumen membrane	wild type leader sequence and μ chain transmembrane domain at C-terminal: NLWTTASTFIVLFLLSLFYSTTVTLF	7
plasma membrane	wild type leader sequence and mutated μ[*] chain transmembrane domain at C-terminal: NLWVVAAVFIVLFLLSLFYSTTVTLF TMD[d] of T cell receptor, met receptor and PDGF receptor	7, 8, 9, 10

[a] PKKKRKV:Nuclear localization sequence of Large T-antigen of SV40 virus.
[b] Aminoterminal presequence of the mitochondrial enzyme cytochrome C oxidase.
[c] Carboxy-terminal tetrapeptide sequence sufficient to cause retention in the ER.[2]
[d] Transmembrane domain.

sequences from proteins secreted in those species have been more efficiently used. The intracellular distribution of a secreted ScFv fragment is illustrated in Figure 7.1D.

Introduction of signals at the C-terminus of antibodies that carry a secretory leader sequence allows to control their localization to the ER lumen, ER lumen membrane or plasma membrane.

ER-Retained Soluble Antibodies

This may be achieved by adding the ER-retaining sequence SEKDEL[2] (see chapter 5) at the C-terminus of the antibody. Antibodies retained in the ER can therefore be used as intracellular anchors to prevent the appearance of a protein on the plasma membrane, or to inhibit the secretion of a protein. Whether this will occur

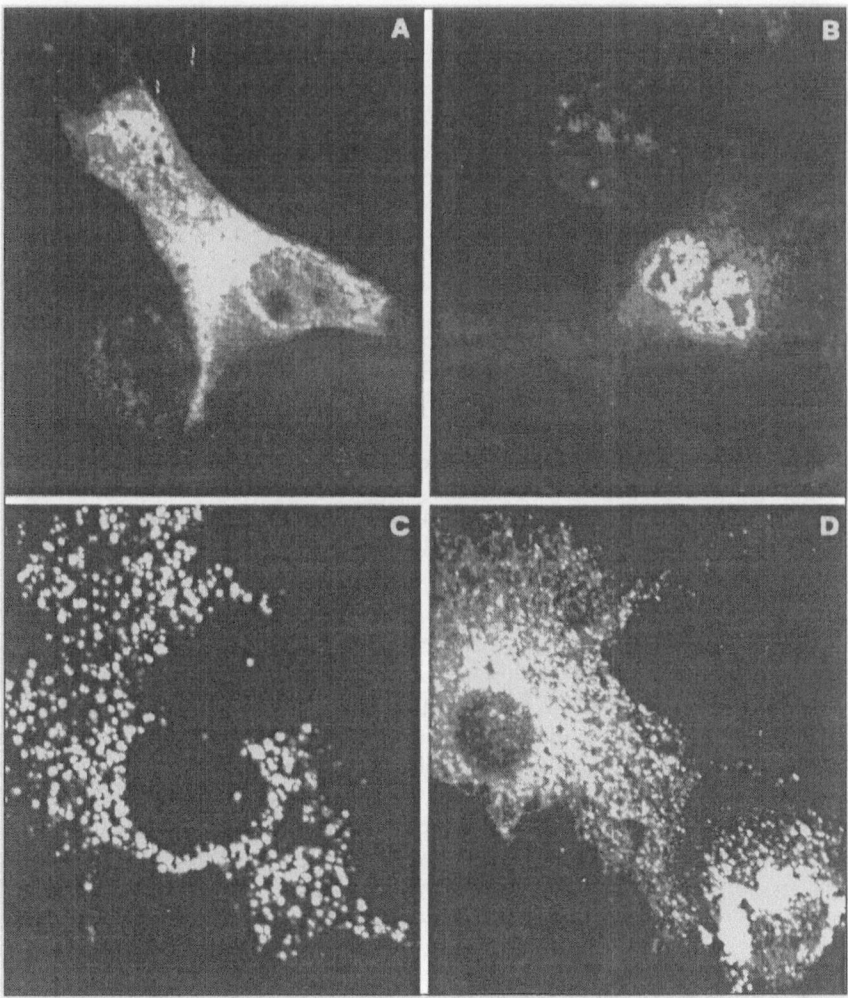

Fig. 7.1. ScFv antibody fragments targeted to different intracellular compartments of COS cells, viewed by immunofluorescence and confocal microscopy, after labeling with fluorescent antibody against a peptide tag at the C-terminus of the ScFv polypeptide. ScFvs were equipped with the following targeting signals: cytoplasmic, (A); nuclear, (B); mitochondrial, (C) and secretory, (D). Reprinted with permission from ref. 3.

will depend on whether the antigen is processed in the ER in a form recognized by the antibody, but this should be verified for each case. The SEKDEL-based retention system is ubiquitous as it is found in all eukaryotic cells. This sequence tag has been added at the C-terminus of ScFv fragments with different specificities and the retention of the corresponding polypeptide in the lumen of the ER has been demonstrated in different mammalian cell lines,[3-5] although in some cases, retention of ScFv fragments was observed also in the absence of SEKDEL signal (see below).

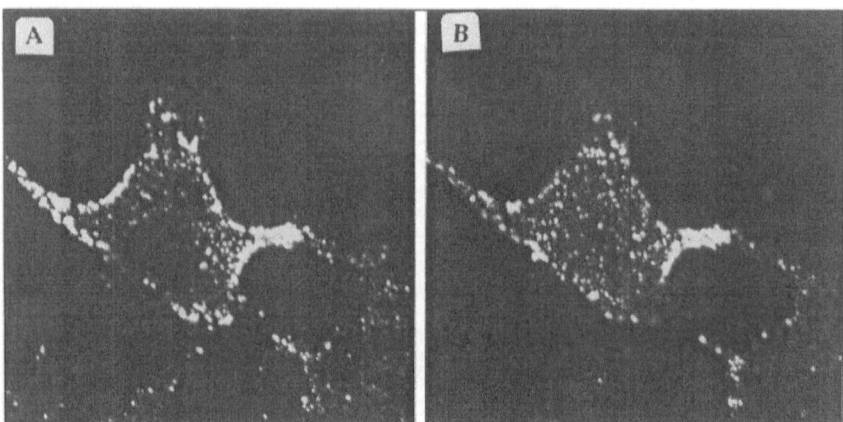

Fig. 7.2. ScFv antibody fragment targeted to the plasma membrane, viewed by immunofluorescence and confocal microscopy. A and B are two focal planes of the same cell. Note the absence of fluorescence into the nucleus in section A and the presence of labeling in the plane of the plasma membrane.

The SEKDEL sequence was also added to the C-terminus of antibody heavy or light chains.[6] Interestingly, in this case, it was found that this signal, although very efficient in inducing retention of assembled immunoglobulins in fibroblasts or glioma derived cell lines, fails to inhibit the secretion of whole antibody molecules from plasma cells. This was not due to a proteolytic cleavage of the C-terminal signal or to the lack of the SEKDEL receptor in these cells as a secreted light chain is efficiently retained by the SEKDEL tag. Rather, it appears that plasma cells are able to override the SEKDEL retention system in order to achieve the secretion of an assembled antibody, whilst efficiently retaining an unassembled antibody chain.

Membrane Antibodies

B cells antibodies also exist as membrane proteins (IgMs) that act as antigen receptors on the cell surface. The transmembrane region of IgM could, in principle, be used to target antibodies to the plasma membrane also in nonlymphoid cells. However, in nonlymphoid cells, the membrane form of IgM is retained in the ER and is not transported to the plasma membrane.[7] Substitution of few amino acids in the µ transmembrane segment, to increase the hydrophobicity of this domain, relieves the intracellular retention and allows IgM to be transported to the plasma membrane in nonlymphoid cells as well.[7] The native µ transmembrane segment can therefore be used to retain antibodies in ER membranes, facing the ER lumen and thus, could be used as an intracellular anchor, distinct from the ER-soluble one.

ScFv fragments have been targeted to the plasma membrane by incorporating the transmembrane and cytoplasmic domains of the T cell receptor ζ chain,[8] the PDGF receptor transmembrane domain[9] and the Met receptor transmembrane domain.[10] An example of this targeting is shown in Figure 7.2.

Table 7.2. Cytoplasmic leaders used for antibody targeting

MGWSCIILFLVATATGVHS	Secretory leader
. . . . S	Mutated leader
. . . . R	(point mutations)
. . . . . . D	
MGWSCKRRSSEETATGVHS	Cytoplasmic leader (hydrophilic mutation)
MA	Cytoplasmic leader (leader-less)

Cytoplasmic Signals

To direct the antibody chains to the cell cytoplasm, the hydrophobic core of their N-terminal hybrophobic leader sequence for secretion has been either mutated or deleted altogether.[11-12] Initial studies showed that substitution of one single hydrophobic amino acid from the secretory leader with hydrophilic residues is only partially effective in targeting the corresponding chains to the cytoplasm with a consistent pool remaining in the ER (SB and AC unpublished). This is due to the corresponding mRNA being part engaged in membrane bound polysomes and part in free polysomes.[13] Mutating most of the hybrophobic core amino acids (6 to 7 residues) into an equal number of charged residues, leads to the complete cytoplasmic localization of the antibody chains (Table 7.2). The complete removal of the leader (leader-less), leaving the N-terminal methionine, also leads to the localization of the corresponding protein to the cytoplasm (Fig. 7.1A) and most likely represents the best choice due to a greater stability of the corresponding protein[14-19] (see below).

Nuclear Signals

The nuclear localization signal PKKKRKV of the large T antigen of theSV40 virus[20] has been initially used to demonstrate that cytoplasmic antibodies could be targeted to the nucleus.[11] The hydrophobic core of the V region leader sequence of the light chain was substituted by this signal and three glycine residues were introduced downstream of the nuclear localization sequence as a spacer to ensure exposure of the nuclear leader from the folded molecule. The resulting polypeptide was synthetized in the cytoplasm and translocated to the nucleus. When the light chain was cotransfected with the wild-type heavy chain, the correct assembly and reconstitution of functional antibody molecules was achieved.[11] More recently efficient nuclear targeting of ScFv fragments has been obtained by adding the NLS of SV40 either at the N- or C-terminal of the molecule[3,17,21] (Fig. 7.1B). In general, C-terminal addition of the NLS tag may be preferable as the potential risk of the NLS hindering the N-terminal antigen binding site would be avoided.

Mitochondrial Signals

The mitochondrial targeting signals (MTSs) are N-terminal extensions (presequences) present in most of the nuclear-encoded mitochondrial proteins (see chapter 5). Similarly to the secretory leader sequence and unlike the NLS, these sequences are removed once the protein is translocated through the mitochondrial membrane. A chimeric protein was made with the N-terminal presequence of the subunit VII of human cytochrome oxidase (COX8.21), covering the cleavage junction, fused to the ScFv fragment. The resulting molecule is made of 25 amino acids of the presequence and the first 10 amino acids of the mature human cytochrome oxidase followed by the sequence of the ScFv and correctly localizes to mitochondria[3] (Fig. 7.1C). The successful targeting to mitochondria expands the range of subcellular compartments where antibody domains can be targeted and is particularly noteworthy as (i) the antibody molecules synthetized are concentrated in the small volume of the organelle and (ii) this could represent the best way to specifically compete with the function of mitochondrial antigens, or with the pool of a given protein that may be found in the mitochondria alongside with other subcellular compartments.

In conclusion, antibody chains and antibody domains can be targeted to a wide spectrum of intracellular compartments and more refined targetings are in principle possible. Evidence about the assembly, folding and antigen binding activity of retargeted antibody chains is described in the following paragraph.

Assembly, Folding and Stability of Intracellular Antibodies

Assembly of antibody chains to form a functional immunoglobulin is a complex process requiring many post-translational events: does it still occur in a cellular context very different from the ER of lymphoid cells such as the cytoplasm? This crucial question was first studied by Carlson[12] in yeast cells and by Biocca et al[11] in mammalian cells, by coexpressing retargeted heavy and light chains. The results of coimmunoprecipitation studies in yeast[12,22] showed that a consistent portion (40-60%) of cytosolic heavy and light chains do indeed associate to form a functional antibody. However, these studies did not formally exclude the possibility that the two chains associate in vitro after the extraction procedure. To assay the association of heavy and light chains in situ, anti-idiotypic antibodies that recognize idiotypes formed by the association of the heavy and light chain variable regions of the parental antibody were used. The results of these studies demonstrate that coexpression of a heavy and a light chain with cytoplasmic localization leads to the formation of stable V_H-V_L complexes in mammalian cells, while coexpression of a secretory antibody chain with the cognate cytoplasmic one does not. A further demonstration of V_H-V_L association in situ was obtained by the coexpression of a "cytosolic" heavy and a "nuclear" light chain. In this case the resulting antibody (as recognized by anti-idiotypic antibodies and by fluorescent antigen) was located in the nucleus, demonstrating that association of the two chains occurred in the cytosol and that nuclear transport of the whole antibody occurred by virtue of the targeting signal present on the light chain alone.[11]

The half-life of the an antibody chain depends very much on the intracellular compartment it is located in. In particular, antibodies and antibody domains expressed in the ER, as secreted or retained proteins, are more stable. On the contrary, the half-life of antibodies in the cytosol appears to be rather short.[11,22] The hydrophilic leader sequence used in the initial studies turned out to have a composition quite similar to that of the so called PEST sequences found in several short-lived proteins[23] (see chapter 5), which may have contributed to the very short half-life. The leaderless format for cytoplasmic expression currently used, certainly improves the half-life of the resulting protein, but it is safe to conclude that cytoplasmic expression is a "worst case" for retargeted antibodies in terms of half-life with respect to other compartments.

There is evidence that cytosolic whole antibody chains are proteolitically cleaved to yield antigen binding, noncovalently associated, Fv fragments.[11] As the covalently linked variable regions (ScFv fragments) are at least 10-fold more stable than the corresponding Fv fragments,[24] the ScFv format appears to be more suitable than whole antibodies for cytosolic expression.

A systematic comparison of the targeting of ScFv fragments to different subcellular compartments, showed that the expression levels of the retargeted antibody domains vary according to the targeting signal they harbor, from high expression levels for those fragments which are synthetized in the ER, to much lower levels for the cytosolic ScFvs with the nuclear and mitochondrial ScFv displaying intermediate levels. This finding most likely reflects the particular properties of the different subcellular compartments with respect to assistance with the folding of antibody domains.[3]

The ER is physiologically involved in the biosynthesis of immunoglobulins and in nonlymphoid cells, it efficiently supports the biosynthesis of whole immunoglobulins.[25] For whole antibodies, the efficiency of their secretion depends on the cell type[25] and appears to be rather independent on the particular antibody. Although secretory ScFv fragments are usually expressed at high levels, the efficiency of their secretion varies.[5] Lack of secretion of ScFvs in the absence of added retention signals has been reported.[26-28] This apparent contradiction may be due to the individual variable regions and/or to cell type specificity. Cryptic retention signals such as the one present in the variable region of some light chains,[29] may play a role in modulating the secretion efficiency of ScFv fragments.

The mitochondrial ScFv expression level is relatively high, comparable to those of its secretory counterpart. As the mitochondrial presequences are cleaved when the proteins are translocated through the mitochondrial membrane, the two forms (immature and mature) should initially be present during the biosynthesis. In the case of the mitochondrial ScFv, the intensity ratio between the immature and the mature forms is of 2:1 after three hours chase. The ScFv protein targeted to mitochondria (mature form) is correctly localized and folded as it retains antigen binding capacity and its intrachain disulfide bonds are formed (see below).[3]

The levels of the cytosolic ScFv fragments are much lower than those of their secretory counterparts expressed from an identical vector. In spite of this fact, cytosolic antibodies have been shown to effectively compete with their target antigens. Understanding the reasons for this limitation will lead to more efficient strategies for intracellular immunization against cytosolic antigens. It is possible that cytosolic ScFv molecules fold poorly, which may lead in turn to a decreased stability of the cytosolic protein. The formation of the intrachain disulfide bonds in the immunoglobulin variable (and constant) domains is a hallmark of Ig domains and is normally catalyzed by specific proteins in the ER (see chapter 4). Antibodies are generally unstable in the absence of disulfide bonds.[24] The redox state of ScFv fragments targeted to the secretory compartment, the mitochondria and the cytosol was compared.[3] While ScFv fragments targeted to mitochondria, as well as secretory ScFv are expressed in their oxidized state in vivo, cytoplasmic ScFv fragments are expressed in a reduced form. This was to be expected, given that the redox potential of the cytoplasm, as measured by the ratio of reduced glutathione to glutathione disulfide (GSH/GSSG) is much higher than that of the secretory compartment.[30] Thus, it is possible that failure to form the intrachain disulfide bonds is the reason for the shorter half-life of the cytoplasmic ScFv fragments, making the partially unfolded antibodies more sensitive to proteolytic attack. However, the working examples reported so far show that folding in the cytoplasm is good enough to preserve antigen binding in most cases. It is worth noting that the presence of the antigen inside the cell appears to stabilize the intracellular antibody (our unpublished observation) in keeping with early in vitro experiments on the refolding of antibody chains in the presence of antigen.[31]

It is noteworthy that in one naturally occurring antibody,[32] one cysteine in the variable region of the heavy chain is absent, suggesting that a disulfide bond in V_H is not necessary for the function of this antibody and may not be obligatory for antibody function in general, at least in the presence of intracellular antigen, despite the almost absolute evolutionary conservation of the variable region cysteine residues. It is also possible that crucial residues in the variable region may play an important role for the folding in ectopic environments, as has been demonstrated for *E. coli* expression.[33]

Another direction for improving the stability of cytosolic antibodies may exploit the information that is being gathered about the control of protein degradation (see chapter 5). In particular, concepts such as the N-end rule, sequences for ubiquitination and destruction boxes should be exploited to improve the stability of cytosolic ScFvs. Thus, the removal of cryptic destruction boxes or ubiquitination signals or the inclusion of signals conferring longer half-life may lead to substantial improvements.

Applications of Intracellular Antibodies

The Yeast System

The intracellular immunization approach has been successfully applied to inhibit the function of several intracellular gene products in different biological systems from yeast to human.[1,34] Studies in yeast have focused on using intracellular antibodies expression to inhibit enzymatic activities related to cell metabolism. In his seminal paper,[12] Carlson introduced into yeast cells leader-less heavy and light antibody chains, neutralizing the activity of the cytoplasmic enzyme alcohol dehydrogenase (ADH I) whose activity can be selected either for or against. To test whether the antibodies neutralized ADH activity intracellularly, cells were plated under conditions that favor the growth of cells containing reduced levels of ADH. Evidence was found that the antibodies did in fact produce a limited degree of neutralization in vivo. It was surmised that the efficiency of neutralization might be improved by genetic selection.

Conversely, in subsequent studies, Hilvert and co-workers exploited a catalytic antibody with chorismate mutase activity, catalyzing a step in aromatic amino acid synthesis which is normally catalyzed by the product of the yeast ARO7 gene.[22,35] The intracellular Fab version of the catalytic antibody was expressed in *aro7* mutant yeast cells, lacking the natural enzyme in the attempt to complement the defect. However, antibody-dependent enzymatic activity was not detected in vivo in these cells, perhaps because the catalytic antibody is 10^4 less efficient that the natural enzyme and was not correspondingly overexpressed. A genetic selection was successfully used to isolate a mutant strain of yeast in which antibody function could be detected in vivo. The antibody was shown to complement the ARO7 defect in this strain, allowing the cells to grow under selective conditions. The mutation did not increase the enzymatic activity of the antibody directly; a possible explanation was suggested that the mutation improved the intracellular expression and/or stability of the antibody even if this hypothesis remained to be determined. Nevertheless, the concept of selecting for cells supporting improved intracellular antibody expression is an important one, likely to find applications in the near future. In chapter 10 we shall extensively discuss the support that the new phage technology can provide to this concept. It is unlikely, however, that the intracellular antibody strategy will supersede genetic techniques in yeast as a means to knock out the function of a gene product in these cells.

The Xenopus Oocytes System and the p21^{ras} Example: A Proof of Principle

For the intracellular immunization strategy to be effective and of practical use, it is necessary to demonstrate that intracellularly expressed antibodies interact in vivo with the corresponding antigen and that this interaction leads to a biological effect. In order to rapidly and effectively verify this in every case of interest, and quite independently of the particular biological system for which the intracellular immunization is planned, we have developed a transient expression system based on *Xenopus laevis* oocytes. In particular, this system has been instrumental to obtaining the first demonstration of intracellular immunization in vertebrate cells by the expression of antibodies retargeted to the cytosol, a "worst case", as discussed above. The mechanistic studies performed on the intracellular expression of antibodies in this system have allowed to gain insight into the strengths of the approach and into the directions that require substantial improvements. The advantages of this system include: (i) oocytes can be easily microinjected with DNA or mRNA for both the antigen and the antibody; (ii) the expression of intracellular antibodies in oocytes is more efficient than in other transient systems tested so far, a point of great importance particularly for cytoplasmic antibodies in light of the above considerations; (iii) the colocalization of the antibodies and their antigens can be studied by fluorescence confocal microscopy; (iv) possibility of performing detailed biochemical studies (redox state, targeting hierarchies etc.).

In many cases the oocyte system also offers a well characterized biological function to study, as in the case of p21^{ras}, a guanine-nucleotide binding protein that is strategically involved in signal transduction process controlling cell growth and differentiation.[36] Moreover, the protein p21^{ras} represents a very important target for cancer therapy since this protein is abnormally activated in a very high percentage of human tumors. Thus, *Xenopus* oocytes can be induced to mature meiotically in vitro by incubation with insulin and the protein p21^{ras} is involved in this process.[37,38] Following the activation of the p21^{ras} protein, oocyte meiotic maturation requires the further activation of the so-called maturation promoting factor (MPF) complex, which leads to germinal vesicle breakdown (GVBD). This active MPF complex contains the serine threonine p34^{cdc2} kinase (also termed histone H1 kinase) whose activity can be readily assayed in oocyte lysates.[39] The genes of the variable regions of the antibody Y13-259[40] were cloned[41] and intracellularly expressed as whole antibodies or as single-chain Fv fragments. The intracellular antibody chains of anti-p21^{ras} specificity interact with the endogenous p21^{ras} protein and by virtue of this interaction, localize in the same submembrane compartment of the animal pole of the *Xenopus* oocyte.[14] This specific interaction has a biological effect that is shown by the fact that in the oocytes expressing intracellular anti-p21[ras] antibody chains or ScFvs, histone H1 kinase activity (induced by insulin, and requiring p21[ras] activity) is markedly reduced (80%) with respect to those expressing a nonrelevant intracellular antibody or ScFv fragment.[14] The oocyte meiotic maturation (which is dependent upon histone H1 kinase activation) is also significantly inhibited (80%) in oocytes expressing anti-p21[ras] ScFvs. Lower expression levels of the anti-p21[ras] induced a similar inhibition of histone H1 kinase activity, but failed to induce inhibition of meiotic maturation.[15] This shows that intracellular antibody expression allows to dissect the pathway upstream of a complex biological function such as meiotic maturation.

The *Xenopus* system may be exploited to monitor and to optimize the efficacy of intracellular antigen-antibody interactions, before and regardless of the particular final application envisaged. Furthermore, we believe that *Xenopus* oocytes will likewise prove to be an excellent system to help the further development of the intracellular antibody technique, for instance, imaging antigen-antibody interactions in vivo and affinity determination of antigen antibody in situ.

p21[ras] is the prototype protein involved in signal transduction. As other signal transduction proteins are located at the cytoplasmic face of the plasma membrane, it is clear that concentrating the antibody at this particular location will improve the effectiveness of the block by increasing the local concentration of the antibody. This can be achieved by incorporating at the C-terminus of the antibody fragments signals for its myristylation or palmitoylation. The 20 C-terminal amino acids of p21[ras], containing the so called CAAX box, have been successfully used to target cytoplasmic proteins to the cytoplasmic face of the plasma membrane (see chapter 5) and we have used it for ScFv targeting (unpublished results).

The Mammalian System

Following the initial proofs of principle for the intracellular antibody approach, a growing number of successful applications have been published. These are summarized in Table 7.3. This synoptic table shows that following the initial "proof of principle" applications, many successful examples have been published in the past three years, with most applications being directed to a clinical setting or to a biotechnological one.

As immediately obvious from the very early days,[11] and as initially demonstrated by the group of Marasco,[26] an important application of the technology is to target proteins important for human viral diseases, most notably HIV type 1 (see chapter 8). Another important field of clinical application concerns the possibility of exploiting intracellular antibodies as antitumoral reagents targeting either cytoplasmic oncoproteins, as in the example of p21[ras] quoted above, or membrane receptors thought to be involved in tumoral progression. It is noteworthy that based on encouraging in vitro results, two clinical gene therapy protocols have been approved by the United States recombinant DNA Advisory Committee (RAC) in 1995.[42] One study will evaluate the efficacy of reinfusing in HIV-1 patients autologous CD4[+] T cells, transduced ex vivo to express a ScFv fragment against gp120. The second one will study a clinical gene therapy protocol to treat with intracellular antibodies metastatic ovarian cell carcinoma in patients.

Table 7.3. Examples of intracellular immunization

Biological System	Target Antigen	Induced Phenotype
Yeast	ADH [12]	Limited degree of neutralization of ADH activity
	Chorismate mutase [22]	Growth advantage under selective conditions
Xenopus laevis	p21^{ras} [14,15]	Inhibition of ras-dependent meiotic maturation
Mammalian cells	HIV-1 gp120 [26]	Reduced HIV-1 infectivity
	p21^{ras} [43]	Inhibition of TCR-induced signal transduction in a T cell line Inhibition of NGF-induced differentiation in PC12 (AC et al, unpublished)
	HIV-1 tat [17]	Inhibition of TAT-mediated transactivation of HIV-1 LTR
	HIV-1 rev [16,64,65]	Inhibition of viral replication
	HIV-1 RT [18,19,63]	Inhibition of early stages of the viral life cycle
	HIV-1 Integrase [66]	Inhibition of early stages of the viral life cycle
	EGF receptor [27] erbB2 [44,45] IL-2Rα [28]	Down regulation of cell surface receptors
	Envelope protein of Tick-borne Flavivirus [67]	Reduced infectivity
Plant cells	Coat protein of the Artichoke Mottle Crinkle Virus [68]	Inhibition of viral replication
	Tobacco Mosaic Virus [69]	Attenuated symptomatology
	Phytochrome [70]	Aberrant seed germination

A third, very successful field of application is represented by the so called plantibodies, the use of intracellular antibody expression to engineer new useful phenotypes in plants. This will be described in chapter 9.

Inactivation of p21^{ras}

The results obtained in *Xenopus* with the anti-p21^{ras} have paved the way to subsequent studies performed with the same antibody in mammalian systems. Since mutations in p21^{ras} protooncogenes have been detected with high incidence in pancreatic carcinomas, colon carcinomas, nonsmall cell lung as well as in many other forms of cancer with poor prognosis with the current available therapies, anti-ras intracellular antibodies may have a role in the gene therapy of these cancers. Studies in a human T cell line confirmed that signal transduction mediated by p21ras was inhibited by the same ScFv anti-p21^{ras} used in previous studies.[14,15,43] Also, differentiation of PC12 cells by NGF is efficiently inhibited by the cytosolic expression of anti-p21^{ras} ScFvs (AC et al, unpublished). Expression of ScFvs format in transgenic mice in a tissue specific manner, would allow their antitumoral activity to be evaluated by mating these mice with transgenic mice prone to develop tissue specific tumors. A first step in this direction before the ScFv anti-p21^{ras} was available, was performed by producing transgenic mice expressing the cytosolic version of the whole immunoglobulin p21^{ras} under the transcriptional control of the immunoglobulin heavy chain promoter (our unpublished work). In these mice we found that the mRNA coding for the anti-p21^{ras} immunoglobulin transgenes in the spleen was developmentally down-regulated in the postnatal period, unlike control immunoglobulin mRNA, suggesting the occurrence of a cell selection induced by the antiproliferative effect of the anti-p21ras intracellular immunoglobulins. We also found that the proliferation rate of splenocytes from the anti-ras transgenic mice, triggered by lectins, was markedly inhibited. The use of intracellular ScFv fragments will greatly facilitate transgenic studies.

In conclusion, the p21^{ras} protein represents a promising target for intracellular antibody applications to tumor therapy. However, all cells express this protein, and the available recombinant antibody Y13-259 neutralizes the activity of both the wild-type and the mutant ras forms. The challenge for the future is to isolate an antibody able to neutralize selectively, the activity of activated ras mutants.

Downregulation of Membrane Receptors

One obvious and simple application of intracellular antibodies is to block proteins in the exocytic pathway. ScFv fragments targeted to the lumen of the ER (generally equipped with the C-terminus retention sequence SEKDEL) provide an effective mechanism for inhibiting the transport on the plasma membrane or the secretion of a protein. This approach has been successfully used so far to inactivate three cell-surface receptors involved in the regulation of cell growth and differentiation and implicated in human cancer. In one study, a single-chain antibody, which competes with Epidermal Growth Factor (EGF) binding to its receptor, was directed to the secretory compartment and expressed in NIH3T3 cells stably transfected with the human EGF receptor gene. The intracellular expression of this ScFv specifically inhibited receptor activation and cell-growth in soft agar in an autocrine fashion.[27] Two different groups have reported the in vivo inactivation of the ErbB2 receptor tyrosine kinase by the intracellular expression of a ER-retained single-chain antibody.[4,44,45] ErbB2 is a 185 kDa transmembrane protein kinase receptor with extensive homology to the family of EGF receptors. Aberrant expression of this receptor is observed in tumors arising mostly in breasts and ovaries and a direct correlation has been noted between overexpression of this receptor and aggressive tumor growth with reduced patient survival. ER-retained ScFvs resulted in the down regulation of cell-surface receptors leading to a selective growth inhibition of ErbB2 overexpressing tumor cells.[44] When expressed in ErbB2 transformed cells, the ScFv bound to the receptor prevented its transit through the endoplasmic reticulum and resulted in functional inactivation and reversion of transformed phenotype in vitro.[4,45] The same approach has been used to inactivate the α subunit of the receptor of human interleukin 2 (IL-2Rα). This receptor is constitutively upregulated in certain T and B-cell malignancies, representing a natural target for suppressive immunotherapy. Intracellular expression of the ER-retained ScFv directed against the α subunit of this receptor abrogates cell surface expression of IL-Rα in stimulated Jurkat T cells.[28]

These examples prove that expressing ER retained ScFv fragments can be very effective to block the traffic of a membrane receptor to the outer plasma membrane. Whether this will prove to be an effective general anticancer therapeutic strategy, in vivo, remains to be demonstrated. In one case[46] it was shown that an ovarian carcinoma cell line, transplanted intraperitoneally, could be transduced in vivo with the anti-erbB2 ScFv gene, resulting in a prolonged survival of animals compared with controls. Many of the membrane receptors which have been chosen as targets for this experimental approach are overexpressed in tumor cells and therefore may be difficult to neutralize completely. Moreover, the intrinsic nature of cancer as a disease, confers a positive selective advantage to cells that are not reached by the therapeutic antibody gene. This may limit the efficacy of intracellular antibodies as a gene therapy approach for cancer treatment, but represents a difficulty in common with many other gene therapeutic approaches to cancer. It is our prejudice, however, that for cancer treatment intracellular antibodies may be more difficult to prove successful in vivo, since their action is confined by the boundaries of each cell; therefore, this requires a 100% targeting efficiency. Rather, intercellular immunization approaches using for instance, gene delivered immunotoxins (see chapter 6), may prove to be more effective.

In this respect, viral diseases stand a better chance as targets for intracellular antibody-based strategies as "immunized" cells may often acquire a selective growth advantage with respected to infected cells (chapters 8, human viral diseases and 9 plant viruses.

Modes of Action of Intracellular Antibodies

In the design of an experiment of intracellular immunization, different aspects of the experimental design have to be taken into account. These will be discussed not only on the basis of what has actually already been done, but also putting forward suggestions that are, in principle, possible with this experimental strategy.

Choice of Promoter

In principle, it is possible to direct the expression of any gene, such as immunoglobulin genes, to any specific cell type by using the regulatory region(s)(promoters and/or enhancers) of a gene that is expressed in that specific cell type. The possibility of achieving transcriptional control of expression of antibodies is an important point of this approach. The crucial issues regarding the choice of promoter for intracellular antibody expression are to accomplish: (i) the highest levels of expression as possible; (ii) the required cell-type specificity and/or (iii) an inducible expression of the antibody. The first point helps to counteract the fact that the half-life of cytosolic immunoglobulins or immunoglobulin domains is shorter than that of their secreted counterparts. In the simplest mode of action, intracellular antibodies act stoichiometrically and therefore their intracellular concentration must match or exceed that of the corresponding antigen. The second and third point is particularly relevant for transgenic studies (see chapter 6). Ideally, inducible systems would not only mediate an "on/off" situation of antibody activity but would also permit to titrate the antibody expression levels (dose response curves). In cultured cells experiments, a good inducible promoter may help to prove formally that an observed phenotype is indeed due to the expression of the intracellular antibody and not to a selection of cells. The inducible promoters available tend to suffer from leakiness of the uninduced state (metallothionein promoter) or from pleiotropic effects caused by the inducing agents themselves (heat shock promoter, mouse mammary tumor virus, induced by dexamethasone). In search of regulatory systems that do not rely on endogenous control elements, several groups have been working to adapt prokaryotic regulatory systems. The most widely used of such systems is that developed by Gossen and Bujard[47,48] which allows regulation of expression of an individual reporter gene over five orders of magnitude. This system is based on control elements of the tetracycline-resistance operon encoded by Tn10, in which transcription of resistance-mediating genes is negatively regulated by the tetracycline repressor (tetR). As a promising alternative to tetracycline, the use of the insect steroid hormone ecdysone as a potent inducer of gene activation in mammalian cells and transgenic mice has been recently reported.[49]

Choice of Antibody Form

The constant domains of antibody chains perform functions that are neither needed nor necessarily exploited for intracellular immunization. Furthermore, despite evidence that the heavy and light chain of immunoglobulins can functionally associate even in the rather hostile cytosolic environment, the whole antibody format is only to be recommended for intercellular immunization experiments (chapter 6). Antibody engineering methods have produced a number of simpler antibody forms initially expressed in *E. coli* (chapter 3) that have been successfully used for intracellular expression.

The use of simpler antibody forms for intracellular expression, such as ScFv fragments, offers the following advantages: (i) the V_H and V_L come on a single polypeptide; (ii) antibodies will be increasingly isolated as ScFv fragments from large phage libraries; (iii) other effector functions may be tagged onto the ScFv; (iv) ScFv, due to their smaller size, may have a facilitated access to some intracellular compartments and (v) ScFv with different specificities and equipped with suitable targeting signals may be coexpressed in a combined approach against different protein targets, possibly in a bi- or polycistronic format.

A caveat in the use of ScFv fragments is that their affinity for antigen may be lower than that of the parental monoclonal antibody so that what one gains in expression levels may be lost in affinity and or specificity. A direct isolation of ScFvs from phage libraries will override this problem (see chapter 3).

One disadvantage of ScFvs is their monovalency. Antibodies are bivalent molecules, but their bivalency is achieved at the expense of being rather large molecules. Small bivalent, homo- or heterospecific antibody domains have been engineered in bacteria as so called diabodies,[50] miniantibodies[51] or chelate antibodies[52] (CRABs). None of these has been expressed intracellularly in mammalian or plant cells but in principle, the intracellular expression of bivalent antibody domains could increase the binding strength for a given antigen through an avidity effect, and bispecificity, could allow to bind and cross-link two different antigens.

Equipping ScFv with Effector Functions

The use of simple antibody forms provide the additional advantage that effector functions could be easily engineered, tailored for particular intracellular immunization applications, much as the effector functions carried by the whole immunoglobulins needed for their action in the immune system. The choice of the particular effector functions that have, or could be, engineered onto intracellular antibodies depends on the mode of action envisaged for the intracellular immunization strategy.

How might intracellular antibodies be employed?

In the simplest case, antibodies would act as dominant negative inhibitors to cause a loss of function. By linking the variable regions to structural elements of the cell or to suitable trafficking signals, antibodies might serve as intracellular anchors, or they could divert the antigen from its normal location, possibly by retargeting it to degradative compartments (suicide antibodies). Intracellular immunization could be applied to interfere selectively with post-translationally modified subsets of a given target protein or with distinct pools of one protein, functioning in one of several cellular compartments. It is clear that these different modes of action will require tailored engineering of the antibody molecule and the choice of antibody forms becomes crucial. In particular, for antibodies to act as competitive inhibitors, no other element apart from the variable regions themselves and the targeting signal is required. This is the experimental format that has been used so far. Also, the use of the ER retention signal SEKDEL has allowed to create intracellular anchors to block the traffic of proteins to the plasma membrane. For cytosolic antigens, the use of intracellular anchors could be extended by engineering fusions with protein domains with binding activity to cytoskeletal elements (such as microtubule binding proteins, actin binding proteins or others). One could conceive to create a ScFv fusion with cytoskeletal proteins themselves to allow the structural incorporation of the antibodies in the cytoskeleton. Antibodies could also be linked to the cytosolic face of the mitochondrial membrane by using a COOH sequence from the bcl-2 protein.[53]

Interfering with retroviruses is one of the most important and successful applications of intracellular antibodies as discussed in chapter 8. Refining the targeting of antibodies close to the sites where the virally encoded proteins act, will improve the effectiveness of the strategy. In one instance, an antibody fragment could carry an "effector function" enabling the structural incorporation of the antibody itself into the viral particle assembled after the primary infection. In the case of HIV, the viral particle contains some proteins important for the viral life cycle, for instance, the reverse transcriptase, the protease and others. These proteins start acting soon after the infection and the effectiveness of any competing intracellular antibody targeted to one of these proteins would be increased if they could act soon after were incorporated into the viral particle. This could be achieved by fusing ScFv fragments to the p6^{gag} [62],[54] vpr or vpx proteins.[55] Thus, the incorporation of intracellularly expressed antibody fusions into the viral particle would lead to the formation of attenuated viral particles, a new example of an antiviral experimental strategy named capsid-targeted viral inactivation.[56-58]

Targeting to a degradative compartment would be one effector function that could be conferred to intracellularly expressed antibodies to create "suicide antibodies". In this scheme, an antibody targeted to a degradative compartment could be engineered to cause the concomitant transport and degradation of the bound antigen. The transit time towards the degradative compartment should be long enough to allow the interaction between the antibody and the corresponding antigen. For antigens present in the secretory compartment, the targeting of antibodies to the lysosomes could be performed either by exploiting the transmembrane and cytoplasmic tail of Lamp 1,[59] or by a fusion with lumenal lysosomal proteins such as cathepsin D. For cytosolic proteins, targeting to the proteasomes could be achieved by fusion with noncleavable ubiquitin, which has been reported to function as a degradation signal when added to heterologous proteins.[60] Ideally, one would require a degradation signal which acts in trans upon binding to the antigen and conditional by an external signal. Although this is at the moment out of sight, the use of "destruction boxes", such as that responsible for the proteolysis of mitotic cyclins[61,62] would be one possibility. Along these lines, another strategy could be to use a bispecific antibody to recruit the ubiquitin-conjugating enzyme(s) to the target antigen in a way similar to that by which this system normally recruits the proteins to be ubiquitinated and degraded. If this was possible, the result would be an antibody-mediated increase of the processive poly-ubiquitination of the target antigen. This would then be targeted for degradation while the antibody, possibly, would not act as a regulatory subunit of the ubiquitinating enzyme.

Antibodies targeted to the plasma membrane by fusion of the ScFv fragment with transmembrane and cytoplasmic domains of receptors and signaling membrane proteins, have been linked to a signal transduction pathway which is activated by a soluble[10] or a cell-bound ligand.[8] In this case, the effector function provided by the chimeric construction is the ability to participate in a signal transduction event controlled by the corresponding antigen. In another realization, a membrane ScFv, fused to the transmembrane and cytoplasmic domain of a rapidly recycling receptor, could be conceivably used as a scavenger of an extracellular ligand, or virus.

Antibodies usually act stoichiometrically with respect to the antigen, and this poses some limitation to intracellular immunization schemes. The use of catalytic antibodies for intracellular expression, coupled to metabolic selection schemes, is an attractive theoretical development, and has been attempted in yeast.[35] However, the general effectiveness of artificial catalytic antibodies is still to be proven. As in the hypothetical example described above, one could conceive the use of intracellular antibodies as a way to target a modifying enzyme (kinase, protease, nuclease etc.) close to the antigen. Many enzymes naturally operate with a binding event by a regulatory subunit or domain. The implementation of such a scheme for antibodies could be done either by direct fusion to an enzyme, or by a bispecific bridging antibody and would confer the antibodies themselves processive properties, with a mode of inhibition that would not be dependent on a continuous binding.

References

1. Biocca S, Cattaneo A. Intracellular immunization: antibody targeting to subcellular compartments. Trends Cell Biol 1995; 5:248-252.
2. Munro S, Pelham HRB. A C-terminal signal prevents secretion of luminal ER proteins. Cell 1987;48:899-907.
3. Biocca S, Ruberti F, Tafani M et al. Redox state of single chain Fv fragments targeted to the endoplasmic reticulum, cytosol and mitochondria. Biotechnology 1995; 13:1110-1115.

4. Graus-Porta D, Beerli RR, Hynes NE. Single-chain antibody-mediated intracellular retention of erb-2 impairs neu differentiation factor and epidermal growth factor signaling. Mol Cell Biol 1995; 15:1182-1191.

5. Jost CR, Kurucz I, Jacobus CM et al. Mammalian expression and secretion of functional single-chain Fv molecules. J Biol Chem 1994; 269:26267-26273.

6. Biocca S, Tafani M, Cattaneo A. Assembled IgG molecules are exported from the endoplasmic reticulum in myeloma cells despite the retention signal SEKDEL. FEBS Letter; submitted.

7. Williams GT, Ventikaraman AR, Gilmore DJ et al. The sequence of the µ transmembrane segment determines the tissue specificity of the transport of immunoglobulin M to the cell surface. J Exp Med 1990; 171:947-952.

8. Eshar Z, Waks T, Gross G, Schindler DG. Specificactivation and targeting of cytotoxic lymphocytes through chimeric single chains consisting of antibody-binding domains and the γ or ζ subunits of the immunoglobulin and T-cell receptors. Proc Natl Acad Sci USA 1993; 90:720-724.

9. Chesnut JD, Baytan AR, Russel M et al. Selective isolation of transiently transfected cells from a mammalian cell population with vectors expressing a membrane anchored single-chain antibody. J Immunol Meth 1996; 193:17-27.

10. Heinrichs AAJ. Antibody display and selection on mammalian cells. PhD thesis, University of Cambridge 1997.

11. Biocca S, Neuberger MS, Cattaneo A. Expression and targeting of intracellular antibodies in mammalian cells. EMBO J 1990; 1:101-108.

12. Carlson JR. A new means of inducibly inactivating a cellular protein. Mol Cell Biol 1988; 8:2638-2646.

13. Mason JO, Williams GT, Neuberger MS. The half-life of immunoglobulin mRNA increases during B-cell differentiation: a possible role for targeting to membranebound polysomes. Genes Dev 1988; 2:1003-1011.

14. Biocca S, Pierandrei-Amaldi P, Cattaneo A. Intracellular expression of anti-p21ras single chain Fv fragments inhibits meiotic maturation of *Xenopus* oocytes. Biochem Biophis Res Commun 1993; 197:422-427.

15. Biocca S, Pierandrei-Amaldi P, Campioni N et al. Intracellular immunization with cytosolic recombinant antibodies. Biotechnology 1994; 12:396-399.

16. Duan L, Bagasra O, Laughlin MA et al. Potent inhibition of human immunodeficiency virus type 1 replication by an intracellular anti-Rev single-chain antibody. Proc Natl Acad Sci USA 1994; 91:5075-5079.

17. Mhashilkar AM, Bagley J, Chen SY et al. Inhibition of HIV-1 Tat-mediated LTR transactivation and HIV-1 infection by anti-Tat single-chain intrabodies. EMBO J 1995; 14:1542-1551.

18. Maciejewski JP, Weichold FF, Young NS et al. Intracellular expression of antibody fragments directed against HIV reverse transcriptase prevents HIV infection in vitro. Nature Medicine 1995; 1:667-673.

19. Gargano N, Cattaneo A. Inhibition of Murine Moloney Leukaemia virus retrotranscription by the intracellular expression of a phage derived anti reverse transcriptase antibody. J Gen Virol 1997; in press.

20. Kalderon D, Roberts BL, Richardson WD et al. A short amino acid sequence able to specify nuclear localization. Cell 1984; 39:499-509.

21. Persic L, Righi M, Roberts et al. Targeting vectors for intracellular immunization. Gene 1997; 187:1-8.

22. Bowdish K, Tang Y, Hicks JB et al. Yeast expression of a catalytic antibody with chorismate mutase activity. J Biol Chem 1991; 266:11901-11908.

23. Rechsteiner M, Rogers SW. PEST sequences and regulation by proteolysis. TIBS, 1996; 21:267-271.

24. Glockshuber R, Schmidt T, Plueckthun A. The disulfide bonds in antibody variable domains: effects on stability, folding in vitro and functional expression in *Escherichia coli*. Biochemistry 1992; 31:1270-1279.

25. Cattaneo A, Neuberger MS. Polymeric immunoglobulin M is secreted by transfectants of nonlymphoid cells in the absence of immunoglobulin J chain. EMBO J 1987; 6:2753-2758.

26. Marasco WA, Haseltine WA, Chen SY. Design, intracellular expression, and activity of a human anti-human immunodeficiency virus type 1 gp120 single-chain antibody. Proc Natl Acad Sci USA 1993; 90:7889-7893.

27. Beerli RR, Wels W, and Hynes NE. Autocrine inhibition of the epidermal growth factor receptor by intracellular expression of a single-chain antibody. Biochem Biophys Res Comm 1994; 204:666-672.

28. Richardson JH, Sodroski JG, Waldmann TA et al. Phenotypic knockout of the high-affinity human interleukin 2 receptor by intracellular single-chain antibodies against the α subunit of the receptor. Proc Natl Acad Sci USA 1995; 92:3137-3141.

29. Dul JL, Argon Y. A single amino acid substitution in the variable region of the light chain specifically blocks immunoglobulin secretion. Proc Natl Acad Sci USA 1990; 87:8135-8139.

30. Hwang C, Sinskey A, Lodish HF. Oxidized redox state of glutathione in the endoplasmic reticulum. Science 1992; 257:1496-1502.

31. Kabat EA. Structural Concepts in Immunology and Immunochemistry. New York: Holt, Reinhart and Winston, Inc, 1968; 157-161.

32. Rudikoff S, Pumphrey JG. Functional antibody lacking a variable-region disulfide bridge. Proc Natl Acad Sci USA 1986; 83:7875-7878.

33. Knappik A, Plueckthun A. Engineered turns of a recombinant antibody improve its in vivo folding. Protein Engineering 1995; 8:81-89.

34. Richardson JH, and Marasco WA. Intracellular antibodies: development and therapeutic potential. Tibtech 1995; 13:306-310.

35. Tang Y, Hicks JB, Hilvert D. In vivo catalysis of a metabolically essential reaction by an antibody. Proc Natl Acad Sci USA 1991; 88:8784-8786.

36. Grand JA, Owen D. The biochemistry of ras p21. Biochem J 1991; 279:609-631.

37. Birchmeier C, Broek D, Wigler M. Ras proteins can induce meiosis in *Xenopus laevis*. Cell 1988; 43:615-621.

38. Deshpande AK, Kung HF. Insulin induction of *Xenopus* laevis oocyte maturation is inhibited by monoclonal antibodies against p21ras. Mol Cell Biol 1987; 7:1285-1288.

39. Arion DL, Meijer R, Brizuela L et al. cdc^2 is a component of the M phase-specific histone H1 kinase: evidence for identity with MPF. Cell 1988; 55:371-378.

40. Furth ME, Davis LJ, Fleundelys B et al. Monoclonal antibodies to the p21 products of the transforming gene of Harvey murine sarcoma virus and of the cellular ras gene family. J Virol 1982; 43:294-304.

41. Werge TM, Biocca S, Cattaneo A. Intracellular immunization: cloning and intracellular expression of a monoclonal antibody to the p21ras protein. FEBS Lett 1991; 274:193-198.

42. Marasco WA. Intrabodies: turning the humoral immune system outside in for intracellular immunization. Gene Therapy 1997; 4:11-15.

43. Werge TM, Baldari CT, Telford JL. Intracellular single chain Fv antibody inhibits Ras activity in T-cell antigen receptor stimulated Jurkat cells. FEBS Lett 1994; 351:393-396.

44. Deshane J, Loechel F, Conry RM et al. Intracellular single-chain antibody directed against erbB2 down-regulates cell surface erbB2 and exhibits a selective anti-proliferative effect in erbB2 overexpressing cell lines. Gene Therapy 1994; 1:332-337.

45. Beerli RR, Wels W, and Hynes NE. Intracellular expression of single chain antibodies reverts erb-2 transformation. J Biol Chem 1994; 269:23931-23936.
46. Deshane J et al. Targeted tumor killing via an intracellular antibody against erbB-2. J Clin Invest 1995; 96:2980-2989.
47. Gossen M, Bujard H. Tight control of gene expression in mammalian cells by tetracycline-responsive pormoters. Proc Natl Acad Sci USA 1992; 89:5547-5551.
48. Gossen M, Freundlieb S, Bender G et al. Transcriptional activation by tetracyclines in mammalian cells. Science 1995; 268:1766-1769.
49. No D, Yao TP, Evans RM. Ecdysone-inducible gene expression in mammalian cells and transgenic mice. Proc Natl Acad Sci USA 1996; 93:3346-3351.
50. Holliger P, Prospero T, Winter G. "Diabodies": small bivalent and bispecific antibody fragments. Proc Natl Acad Sci USA 1993; 90:6444-6448.
51. Pack P, Plueckthun A. Miniantibodies: use of amphipathic helices to produce functional, flexibly linked dimeric Fv fragments with high avidity in *E. coli*. Biochemistry 1992; 31:1579-1584.
52. Neri D, Momo M, Prospero T et al. High-affinity antigen binding by chelating recombinant antibodies (CRAbs). J Mol Biol 1995; 246:367-373.
53. Nguyen M, Millar DG, Yong VW et al. Targeting of Bcl-2 to the mitochondrial outer membrane by a COOH-terminal signal anchor sequence. J Biol Chem 1993; 268:25265-25268.
54. Kondo E, Mammano F, Cohen EA et al. The $p6^{gag}$ domain of human immunodeficiency virus type 1 is sufficient for the incorporation of Vpr into heterologous viral particles. J Virol 1995; 69:2759-2764.
55. Wu X, Liu H, Xiao H et al. Targeting foreign proteins to human immunodeficiency virus particles via fusion with Vpr and Vpx. J Virol 1995; 69:3389-3398.
56. Natsoulis G, Boeke JD. New antiviral strategy using capsid-nuclease fusion protein. Nature 1991; 352:632-635.
57. Natsoulis G, Seshaiah P, Federspiel MJ et al. Targeting of a nuclease to murine leukemia virus capsids inhibits viral multiplication. Proc Natl Acad Sci USA 1995; 92:364-368.
58. Schumann G, Qin L, Rein A et al. Therapeutic effect of Gag-nuclease fusion protein on retrovirus-infected cell cultures. J Virol 1996; 70:4329-4337.
59. Roher J, Schweizer A, Russel et al. The targeting of Lamp1 to lysosomes is dependent on the spacing of its cytoplasmic tail tyrosine motif relative to the membrane. J Cell Biol 1992; 11:497-505.
60. Johnson ES, Bartel B, Seufert W et al. Ubiquitin as a degradation signal. EMBO J 1992; 11:497-505.
61. Glotzer M, Murray AW, Kirschner MW. Cyclin is degraded by the ubiquitin pathway. Nature, 1991; 349:132-138.
62. Klotzbucher A, Stewart E, Harrison D et al. The 'destruction box' of cyclin A allows B-type cyclins to be ubiquinated, but not efficiently destroyed. EMBO J, 1996; 15:3053-3064.
63. Shaheen F, Duan L, Zhu M et al. Targeting human immunodeficiency virus type 1 reverse transcriptase by intracellular expression of single-chain variable fragments to inhibit early stages of the viral life cycle. J Virol 1996; 70:3392-3400.
64. Wu Y, Duan L, Zhu M et al. Binding of intracellular anti-rev ingle-chain variable fragments to different epitopes of human immunodeficiency virus type 1 rev: variation in viral inhibition. J Virol 1996; 70:3290-3297.
65. Duan L, Zhang H, Oakes JW et al. Molecular and virological effects of intracellular anti-rev single-chain variable fragments on the expression of various human immunodeficiency virus-1 strains. Human Gene Therapy 1994; 5:1315-1324.

66. Levy-Mintz P, Duan L, Zhang H et al. Intracellular expression of single-chain variable fragments to inhibit early stages of the viral life cycle by targeting human immunodeficiency virus type 1 integrase. J Virol 1996; 70:8821-8832.

67. Jiang W, Venugopal K, Gould EA. Intracellular interference of Tick-Borne Flavivirus infection by using a single-chain antibody fragment delivered by recombinant Sindbis virus. J Virol 1995; 69:1044-1049.

68. Tavladoraki P, Benvenuto E, Trinca et al. Transgenic plants expressing a functional single-chain Fv antibody are specifically protected from virus attack. Nature 1993; 366:469-472.

69. Voss A, Niersbach M, Hain R et al. Reduced virus infectivity in *N. tabacum* secreting a TMV-specific full-size antibody. Mol Breeding 1995; 1:39-50.

70. Owen M, Gandecha A, Cockburn B et al. Synthesis of a functional anti-phytochrome single-chain Fv protein in transgenic tobacco. Bio/Technology 1992; 10:790-794.

CHAPTER 8

Gene Therapy and Research Applications of Intrabodies for Human Infectious Diseases

Wayne A. Marasco

Gene therapy is a new form of molecular medicine that holds tremendous promise for the treatment of human infectious diseases. Perhaps no better example of the broad impact of gene therapy for the treatment of human infectious diseases comes from studies on the human immunodeficiency virus, HIV-1. Gene therapy for the treatment of HIV-1 infection and AIDS has captured the interest of a number of investigators as an attractive addition to conventional pharmacologic therapies because alteration of the host cell could potentially confer permanent suppression of viral replication after infection or even lasting protection against viral infection. Several strategies have been investigated that have different, and sometimes overlapping modes of action. One approach is based on enhancement of the immune response against the virus by using genetically modified cells that express viral gene products to induce antiviral cellular immune responses. Another approach uses an extracellular mode of action by secretion from the transduced cell of a factor directly affecting the HIV-1 life cycle (e.g., sCD4-IgG, anti-gp120 Fab antibody fragments) or of a factor affecting the host's defense mechanism (e.g., a cytokine). Intracellular Immunization is a third strategy that is aimed at the stable transfer of genetic elements that inhibit viral replication, so-called "resistance genes" into those cells of a patient that are potential targets for viral infection.[1-3]

Intracellular antibodies or "intrabodies" are a recent addition to the field of intracellular immunization-based strategies being used to treat infectious and other diseases.[4] As an anti-HIV-1 treatment strategy, these anti-HIV-1 intrabodies are synthesized by the cell and targeted to specific cellular compartments where they bind to their target HIV-1 protein and inhibit its function. In this chapter I will summarize the studies from our laboratory on the use of anti-HIV-1 intrabodies against HIV-1 envelope glycoprotein gp120, Tat, Rev and matrix protein (MA, p17) as research tools and on their applications for gene therapy of HIV-1 infection and AIDS. Other investigators have reported on the use of anti-HIV-1 intrabodies against Rev[5-7] and reverse transcriptase;[8] these studies will be reviewed as well. Other

Intracellular Antibodies: Development and Applications, edited by
Antonino Cattaneo and Silvia Biocca. © Springer-Verlag and Landes Bioscience 1997.

reports that describe antiviral effects of intrabodies against HTLV-1[9] and tick-borne flavivirus[10] will not be reviewed. It will become clear from the HIV-1 studies presented here that intrabodies exhibit tremendous versatility in their ability to inhibit different stages of the viral life cycle by targeting structural, regulatory and enzymatic proteins of the virus.

Inhibition of Postintegration Events of the HIV-1 Life Cycle

Endoplasmic Reticulum (ER)-Expressed Anti-HIV-1 GP120 sFv Intrabodies Disrupt GP160 Processivity and Maturation

The envelope protein of HIV-1 is located on the cell surface of HIV-1 particle. The glycosylated envelope protein precursor gp160 is translocated into the ER after synthesis and is cleaved within the Golgi apparatus to yield a mature envelope protein composed of gp120 and gp41. The human monoclonal antibody F105 that was used in these studies was shown to compete with CD4 for binding to gp120 and represents a class of broadly neutralizing antibodies that are directed against the CD4 binding site.[11-13] This hybridoma was used to engineer a single-chain (sFv) intrabody termed, sFv105.[3,14] The strategy was to design an sFv intrabody that would be directed into the lumen of the ER and be retained in this region of the secretory pathway so that it could bind gp160 precursor and prevent its transport to the cell surface. For these studies, two different sFv105 intrabodies were designed which differed only by the presence of a C-terminal SEKDEL sequence. This sequence is predicted to cause the retention of soluble proteins in the lumen of the ER.[15] Both intrabodies contained the native heavy chain leader sequence to direct their synthesis into the lumen of the ER. Unexpectedly, sFv105 which was designed for secretion was stably retained in the ER and could be coimmunoprecipitated with a rat monoclonal antibody to the ER chaperone protein, BiP (GRP78).[16] In marked contrast, the sFv105KDEL intrabody yielded an unstable protein that was rapidly degraded if not coexpressed with envelope glycoprotein.[3] This unusual behavior of the sFv105 and sFv105KDEL in terms of ER-retention and stability is not a general observation. In fact, in our laboratory, ER-directed intrabodies that contain the KDEL retention sequence and are directed against cellular surface molecules (e.g. Tac, the high affinity α-chain component of the IL-2R receptor[17]) appear to be both more completely retained in the ER and more efficient in their ability to cause complete "phenotypic knock-out" than the same sFv intrabodies that do not contain the ER-retention signal. Interestingly, the KDEL retention signal does appear to influence that intracellular fate of the intrabody/target protein complexes.[17]

Neither intrabody was toxic to the cells when stably expressed. The results of these early studies were very conclusive in that in stably transfected sFv105 expressing cell lines that were infected with HIV-1 there was: (1) marked inhibition of processing of the HIV-1 envelope precursor; (2) a decrease in envelope-mediated syncytia formation and (3) substantial reduction in the infectivity of the HIV-1 particles released by the sFv105 intrabody-producing cells.[3] In addition, in contrast to cells stably transfected with vector alone in which surface CD4 was down-regulated following HIV-1-infection (association of gp160/CD4 in the lumen of the ER is known to cause down-regulation of surface CD4[18-20]) in the sFv105 expressing HIV-1-infected cells, surface gp120 expression was markedly reduced and surface CD4 was normal.[14] These results demonstrated that intracellular protein-pro-

tein interactions could be inhibited by the sFv intrabodies. Cell surface phenotype, replication rate, and response to mitogenic stimulation of the sFv105 expressing cells were also normal.

Transduction of the sFv105 gene into PBMCs from both uninfected and HIV-1-
infected patients was achieved using Murine leukemia virus (MuLV) vectors that
encoded the sFv105 gene under the control of an internal CMV promoter.[21,22] In
HIV-1 challenge experiments, both cytoprotection and inhibition of HIV-1 replication were observed. Based on these encouraging results, a clinical gene therapy
protocol was approved by the United States Federal Recombinant DNA Advisory
Committee (RAC) in June, 1995. This study will evaluate the safety and efficacy of
intracellular antibody gene therapy in asymptomatic patients with HIV-1 infection by reinfusing autologous CD4$^+$ T cells that have been transduced ex vivo with
a MuLV vector that expresses the sFv105 intrabody. The in vivo kinetics and survival of sFv105-transduced cells will be compared by limiting dilution PCR with
those of a separate aliquot of cells transduced with a control vector (identical except for the sFv105 gene). The level and persistence of sFv105 expression and various immunologic parameters including CD4 counts, virus load and cytotoxic T-cell
activity against the transduced cells will also be assessed. The results of these studies should help determine whether this intracellular antibody can protect CD4$^+$ T
cells in patients with HIV-1 infection. The results should also aid in the design of
future trials of larger scale T cell replacement and of hematopoietic stem cell gene
therapies of HIV disease.

Combined Intra- and Extracellular Immunization Using an Anti-HIV-1 GP120 Fab Intrabody that Is Secretion Competent

The strategy of using anti-gp120 intrabodies to inhibit HIV-1 replication was
extended to develop an anti-HIV-1 approach that uses the combined intra- and
extracellular binding activities of neutralizing Fab fragments of F105.[23] This structural modification (using the Fab fragment of F105 instead of the sFv configura-
tion) allowed the Fab105 fragments to be efficiently secreted from the cell which in
turn would allow the Fab105 fragments to interact with gp160/gp120 both in the
secretory compartment and outside the cell. For these studies, separate CMV promoters in the same orientation were used to drive expression of the heavy and
lights chains. Human CD4$^+$ T cells constitutively expressing and secreting the Fab105
fragments were shown to be resistant to HIV-1 infection by intracellular binding
to the envelope glycoprotein. Moreover, the secreted Fab105 fragments were able
to block new rounds of HIV-1 infection. Direct cell-to-cell transmission of the virus by extracellular neutralization of the virus particles and by binding to gp120
on the surface of nontransduced HIV-1-infected CD4$^+$cells, respectively, was also
inhibited. Another remarkable observation that was made was that the nascent
Fab105 intrabody fragments could inhibit infectious HIV-1 production by binding
intracellularly to envelope mutants that escape neutralization by extracellular F105
antibody. These observations suggested that the high concentrations of Fab105 fragments in the secretory vesicles could compensate for decreases in affinity that resulted from point mutations in the envelope glycoprotein. Similar observations
have been reported with lower affinity CD4 mutants that were retained intracellularly.[24] Thus, by using this powerful combined *cis* and *trans* approach, the stably
transfected F105 Fab expressing cells were immunized against HIV-1 infection by
the intracellular binding activity of the nascent Fab fragments and also provided
protection to the surrounding HIV-1 susceptible cells from virus infection.[23]

To practically use this above approach for HIV-1 gene therapy, it is a prerequisite that the targeted cells, such as primary CD4$^+$ lymphocytes/monocytes and CD34$^+$ stem cells, can be efficiently transduced with the antibody gene and that the heavy and light chains are expressed in near stoichiometric amounts. Adeno-associated virus (AAV), a defective, nonpathogenic parvovirus is another suitable vector for HIV-1 gene therapy because AAV is able to incorporate DNA into the cell chromosome independent of DNA replication and, therefore, to transduce the therapeutic gene(s) into replicating as well as nonreplicating cells.[25,26] In addition, AAV-mediated gene transfer would also eliminate the potential problem of activation of HIV-1 during the necessary activation step (to dissolve the nuclear membrane) that is required for MuLV gene transfer because MuLV does not contain a nuclear localization motif in its preintegration complex. [27-30]

To optimize the coexpression of the heavy and light chains of the Fab105 intrabodies, a bicistronic expression vector pCMV-Fab-IRES, under the control of the cytomegalovirus immediate early (CMVIE) promoter, was constructed that allows the near stoichiometric coexpression of the Fd heavy and complete light chain of an Fab fragment.[4,31] An internal ribosomal entry site (IRES) corresponding to the 5' UTR of encephalomyocarditis virus (EMCV)[32,33] was used to obtain CAP-independent ribosomal binding and high level translation of the light chain. The Fab105 expression cassette was then cloned into an adeno-associated virus (AAV) shuttle vector, and encapsidated recombinant AAV-Fab105 vectors were produced. The Fab105 intrabody genes were shown to be transduced into human lymphocytes by using the recombinant AAV viruses and the infection with several primary HIV-1 patient isolates was effectively blocked in the transduced T-lymphocytes.[34]

Cytoplasmic Expression of Anti-HIV-1 Tat sFv Intrabody Inhibits Tat-Mediated HIV-1 LTR Transactivation and Viral Replication

Tat is a small one (72-amino acid) or two (86-amino acid) exon regulatory protein encoded by the HIV-1 genome that plays a unique role in the emergence of the virus from the latent state. Tat is known to be a potent transcriptional activator of the HIV-1 long terminal repeat (LTR).[35-37] The sequence in the 5' untranslated region of all HIV RNAs called transactivation response element (TAR) (nucleotides +1 to +44), is required for the Tat-mediated transactivation. Tat binds with high efficiency to a bulge region in a stable stem-loop structure of the TAR RNA and then interacts with a transcription initiation complex[38,39] composed of Tat, DNA- and RNA-binding proteins[37,40-45] to primarily stimulate transcription initiation and increase transcription elongation.[36,46,47]

Tat has other modes of action as well. Expression of Tat in human cells in culture leads to transactivation and overexpression of cellular genes encoding cytokines such as TNFα, TNFβ and IL-2.[48-51] Some of which in turn can activate viral transcription through their activation of NF-κB.[50,51] NF-κB binding to enhancer elements in the HIV-1 LTR can lead to *tar*-independent activation of viral transcription by Tat.[52,53] Superactivation of HIV-1 LTR-driven gene expression is thus induced by the concerted interaction of cellular trans-activator NF-κB and Tat.[52,53,55,56] Indeed, cooperative inhibition of NF-κB and Tat-induced superactivation of HIV-1 LTR-regulated gene expression has been reported.[55] Moreover, viruses with a defective *tat* gene cannot proliferate and do not show cytopathic effects.[57,58]

Anti-Tat sFv intrabodies were constructed against two regions of Tat protein using exon1- or exon2-specific monoclonal antibodies and tested for inhibition of Tat-mediated transactivation of viral mRNA transcription and viral activity.[59] The coding sequences of all anti-Tat sFvs were modified to contain a Kozak consensus sequence and start methionine immediately 5' to amino acid one of framework one of the heavy chain.[60] For each anti-Tat sFv, an additional anti-Tat sFv derivative (sFv-fusion protein) was constructed for cytoplasmic expression that had the entire human κ chain constant domain ($C_κ$) fused in frame at the carboxy-terminus of the sFv cassette (sFvtat$C_κ$), and was designed to increase the stability and/or possibly promote dimerization in a form similar to Bence Jones proteins.[61] For nuclear targeting, the sFvtat and sFvtat$C_κ$ were additionally modified to contain the carboxy-terminal SV40 nuclear localization signal (sFvtatSV40 and sFvtat$C_κ$SV40, respectively) which has been shown to direct intracellular antibodies and sFv fragments into the nucleus.[62] All anti-Tat sFvs were cloned under the control of the human cytomegalovirus immediate early promoter (CMVIE).

Several important specific and general results were obtained. First, the anti-Tat sFvs with specific binding activity against the exon1 N-terminal activation domain of Tat were able to block Tat-mediated transactivation of HIV-1 LTR (using a CAT reporter gene), as well as, intracellular trafficking of Tat in mammalian cells despite the fact that Tat is present in very low concentrations in HIV-1-infected cells. As a result, the stably transfected lymphocytes expressing anti-Tat sFvs were resistant to HIV-1 infection. An alternative anti-Tat sFv directed against the exon2 C-terminus was unable to suppress HIV-1 replication, indicating that critical importance of the specific targeted epitope. Second, nuclear targeting of the anti-Tat sFv was not required to inhibit the function of Tat (which acts in the nucleus). This result suggested that by inhibiting nuclear import of Tat, Tat/TAR RNA interactions in the nuclear compartment could effectively be inhibited. Third, in stably transfected CD4$^+$ T cells sFvtat$C_κ$ fusion proteins had greater inhibitory activity than their sFvtat counterparts which suggested that the addition of the $C_κ$ domain resulted in increased stability of the sFv intrabody and/or possibly dimerization,[61] and fourth, stably expressed sFv intrabodies and their modified forms could effectively target molecules in the cytoplasm and nuclear compartments of eukaryotic cells and folding of the sFv intrabodies to form a functional binding site could occur in the reducing environment of the cytoplasm. However, the efficiency of this folding process is unknown.

Transduction and expansion studies on CD4$^+$ T-cells derived from HIV-1-infected individuals by MuLV based vectors encoding the most potent N-terminal anti-Tat sFv intrabody, termed sFvtat1$C_κ$, have been performed.[22] These MuLV vectors are similar to the vectors described for the sFv105 studies except for two features. First, the sFvtat1$C_κ$ gene was used in place of the sFv105 gene. Second, a β-galactosidase gene with an intranuclear localization signal was used in place of neomycin to allow for microscopic determination of transduction efficiencies. In these studies, expression of the sFvtat1$C_κ$ intrabody was more effective at stably inhibiting HIV-1 replication in transduced cells from HIV-1-infected individuals than was sFv105.

Cooperative inhibition of NF-κB and Tat-induced superactivation of HIV-1 LTR-regulated gene expression has been reported using pharmacologic agents.[55,63] However, a combined pharmacologic and genetic therapeutic approach to these same targets had not been investigated. Accordingly, studies were initiated to determine if potentiation of the inhibitory activity seen with the sFvtat1C_κ intrabody would occur in the presence of two different NF-κB inhibitors, pentoxifylline (PTX)[63,64] and Gö-6976.[65] These studies showed that in the presence of Gö-6976 alone or in combination with PTX, both stably transfected, sFvtat1C_κ expressing CD4$^+$ Jurkat-T cells and transduced PBMCs, when challenged with both laboratory strains and primary isolated of HIV-1, showed significant prolongation of inhibition of HIV-1 replication when compared to drug treated cells alone or to sFvtat1C_κ intrabody expressing cells alone.[145] This combined effect was more than additive. These studies suggest that combined pharmacologic and genetic based anti-viral regimens that block different steps of the TAR dependent/independent Tat activation pathways may have therapeutic value in the clinical setting of HIV-1-infection.

Cytoplasmic Expression of Anti-HIV-1 Rev sFv Intrabody Inhibits Rev-Mediated HIV-1 LTR Transactivation and Viral Replication

Rev is another small 116 amino acid, 20 kDa HIV-1 regulatory protein that is the regulator of virion protein expression.[66] Rev is encoded by multiple spliced viral mRNA and appears to efficiently function only in multimeric form.[67,68] Rev controls virion protein expression by controlling the cytoplasmic accumulation of RNA species. Rev binds in the nucleus to the cis-acting Rev-responsive element, designed RRE, that is present in unspliced (genomic and gag/pol mRNA) and singly spliced (env mRNA) viral RNA. Rev interacts with cellular cofactors including a nucleoporin protein[69,70] and this allows cytoplasmic transport of unspliced and singly spliced viral mRNA to occur, ultimately allowing HIV-1 replication.[66,71,72] In the absence of an active Rev protein, only spliced mRNAs can be transported to the cytoplasm. The sequences coding for the capsid replicative enzymes and envelope glycoprotein have been removed by splicing.

Duan et al[5] reported that a cytoplasmic sFv intrabody against Rev, when stably transfected into CD4$^+$ HeLa cell lines, inhibited nuclear transport of Rev, syncytia formation and HIV-1-production.[5] HIV-1 infection was also inhibited in human PBMCs that were transduced with retroviral vectors expressing the anti-Rev sFv.[7] Interestingly, no alterations in HIV-1 internalization, reverse transcription, or initial transcription of multiple spliced viral mRNAs were demonstrated despite inhibition of HIV-1 replication.[6] These results imply that the anti-Rev sFv intrabody is working in a manner similar to the anti-Tat sFvs,[59,73] namely, by binding Rev in the cytoplasm and preventing its nuclear translocation and possibly its proposed nucleocytoplasmic shuttling activities.[69-72,74] As a result, subsequent interaction of Rev with cellular Rev binding proteins[69,70,75,76] or with the RRE containing viral mRNA and associated RRE binding proteins[77,78] may be inhibited.

The anti-Rev sFv intrabodies that have been constructed in our laboratory have confirmed, with a second HIV-1 protein, the critical importance of the epitope that is targeted. Both our studies, as well as the results of Wu et al,[79] show that the intrabodies that are directed against the activation domain on Rev have more potent inhibitory activity than sFv intrabodies that are directed against a nonactivation region of the Rev protein (Mhashilkar & Marasco, unpublished observations).

Inhibition of Preintegration Events of the HIV-1 Life Cycle

Intracellular immunization strategies that are focused on the earliest events of the viral life cycle may offer the greatest hope for blocking the infection of susceptible cells (preventing the establishment of infection might be more efficient) than strategies aimed at inhibiting gene expression after integration of viral DNA into the cellular genome. Virus uncoating, reverse transcription, transport of the preintegration complex (containing viral DNA) to the nucleus, and subsequent integration of the viral DNA into host chromosomes, are all necessary steps that must occur for infection to be established. These early events of the viral life cycle are potential steps that could be inhibited using anti-HIV-1 intrabody based strategies.[4] This strategy may allow HIV-1-infection of susceptible cells to be aborted and true "intracellular immunization" may be achieved.

Recent studies have identified components of the preintegration complex to include p17(MA), Vpr, integrase and reverse transcriptase as well as viral genomic RNA.[80-82] Nuclear localization signals (NLS) on p17(MA)[83] and Vpr[84] have also been identified which appear to allow the establishment of HIV-1-infection in macrophages and quiescent T lymphocytes where the nuclear membrane is intact. Anti-HIV-1 intrabodies that are directed against critical epitopes on these proteins, perhaps optimally at nuclear localization signals for MA and Vpr or against the active sites of integrase or reverse transcriptase, may provide another avenue by which these intrabodies may prevent integration by blocking the nuclear transport or disabling the preintegration complex. Likewise, intrabodies against the HIV-1 capsid protein (CA,p24), may disrupt viral uncoating and virus particle formation[85] while intrabodies that are directed against nucleocapsid (NC,p7) may inhibit reverse transcription and packaging of viral genomic RNA.[86-89] In the following section, inhibitory activity against two of these HIV-1 targets will be described.

Inhibition of Early and Late Events of the HIV-1 Replication Cycle by Antimatrix Protein Intrabodies

Matrix protein (MA, p17) is thought to be involved in two critical stages of the viral life cycle as it is required for both nuclear import of the viral preintegration complex and for particle assembly. Following acute infection, MA has been detected in cell fractions containing partially purified HIV-1 preintegration complexes[81] as well as in the nucleus, (the latter due presumably to the presence of a NLS on MA.)[83,90] Mutant viruses that show a defect in nuclear import have an impaired ability to infect nondividing cells such as macrophages.[28,84] In addition to fulfilling this critical function at an early step of the infection process, MA also plays an essential role in virus morphogenesis. A myristate residue and charged N-terminal amino acids of MA directs Gag to the plasma membrane. This targeting is essential for the proper assembly of viral particles[91,92] and for their release into the extracellular space.[93-95] In the process, MA also recruits the envelope glycoprotein at the surface of virions.[96-98] Molecular explanations for these apparently two opposing effects of MA (two subcellular localization signals) have recently been reported. [99-101]

To investigate whether anti-MA intrabodies could inhibit critical early and/or late events in the virus life cycle, we constructed Fab intrabodies that were directed against MA. Coexpression of the Fd and light chains was optimized using the pCMV-Fab-IRES vector (ref. 4 and Levin, submitted). In this construct, the native immunoglobulin leader sequences of the Fd heavy and κ chains were removed, individual chains were modified to contain a Kozak consensus sequence and an ATG initiation codon immediately preceding amino acid one of framework one of the heavy chain and amino acid minus four of the leader sequence of the κ chain.[59] The binding site for the 3H7 MAb used in these studies had been previously epitope mapped to the amino acid sequence KKAQQAAADT (residues 113-122) near the carboxy terminus of MA.[102] This epitope is associated with the Clade B HIV-1 genotype,[103] including well-studied laboratory strains such as LAI, MN, SF2 and RF, as well as a majority of European and North American isolates. Isolates of Clade B may be distributed globally.[103]

For these studies, the preintegration and postintegration events of the virus life cycle were examined separately. To examine the role of the anti-MA intrabodies in the preintegration phase of the viral life cycle, the ability of cytoplasmic Fab3H7 intrabodies to inhibit HIV-1 gene expression was investigated using a HIV-1 CAT virus that has a chloramphenicol acetyltransferase (CAT) gene replacing the *nef* gene and a deletion in the *env* gene. Wild type HIV-1 envelope was supplied in *trans*, so that the virions had the genetic capacity for only a single round of infectivity.[104] Following infection and integration of viral DNA, newly synthesized Tat protein transactivates the expression of the CAT gene. If integration is blocked, Tat-mediated gene expression can still occur presumably from unintegrated viral DNA, however, it is relatively inefficient.[105-108] In general, there is an insufficient level of HIV-1 gene expression to support a spreading viral infection.[105-109] When stably transfected Jurkat-Fab3H7 cells were challenged with the CAT virus, marked inhibition of CAT activity was seen. This inhibition could occur as a result of inhibition of any critical step in the afferent arm of the virus life cycle that leads to integration of the HIV-1 provirus.

To determine if the stable Jurkat-Fab3H7 cells were resistant to HIV-1 infection, stable Jurkat-Fab3H7 cells were challenged with both laboratory strains of HIV-1 and European, syncytium-inducing (SI) primary isolates that had been passaged on activated PBMCs and screened for their ability to directly infect Jurkat cell lines. The results of several experiments demonstrated that there was a marked delay of virus infection (as measured by p24 release) over the 20-25 day experiments in Jurkat-Fab3H7 cells compared to parental Jurkat cells or stable Jurkat-vector cells. In addition, in these experiments were multiple round of infection are occurring and proviral integration were examined directly, integration was delayed but not prevented in the anti-MA Fab intrabody expressing cells. Thus, these experiments demonstrated that actively dividing $CD4^+$ Jurkat T cells stably expressing anti-MA Fab intrabodies are resistant to HIV-1 infection when challenged with both laboratory strains and primary viral isolates. However, this resistance is relative and not absolute with the stable cell lines and HIV-1 strains that were examined.

To examine the role of the anti-MA intrabodies in the postintegration phase of the viral life cycle, the infectivity of the virus particles released from the HIV-1-infected Jurkat-Fab3H7 cells was studied. Marked inhibition of virion infectivity was seen with virions released from the Fab3H7 intrabody expressing cells, compared to control cells with all viruses that were examined. This inhibition of virion infectivity ranged from 76% to 98% and with the majority of subclones showed greater than 90% inhibition compared to the virions released from control cells. Thus, these experiments demonstrated a significant decrease in the infectivity of the virions released from these HIV-1-infected Fab3H7 intrabody expressing cells.

In summary, in Jurkat-Fab3H7 cells challenged with HIV-1 marked inhibition of proviral gene expression occurred when single round HIV-1 CAT virus was used for infections. When multiple rounds of infection using both laboratory strains and syncytium-inducing primary isolates were examined in HIV-1 challenge experiments, a substantial delay in the spread of infection and a reduction in the infectivity of virions released from the cells was observed. The precise mechanism(s) of inhibition of Tat-mediated gene expression or of the inhibition of infectivity of virus particles released from the intrabody expressing cells must still be determined. Further experiments with single round infections will be required to delineate the precise mode(s) of action for the anti-MA Fab intrabodies on the afferent and efferent arms of the replication cycle. Likewise, it is not known whether the "break-through" of virus in the intrabody producing cells is due to the development of mutant resistant viruses. Indeed, this epitope is not invariant even for the B clade[110] and amino acid changes in this region are tolerated and generate infectious virus.[96,111]

For the inhibitory effects of the anti-MA Fab intrabodies to be fully elucidated, further studies will be required in terminally differentiated, nondividing cells such as macrophages (monocytes, tissue macrophages, dendritic cells, and microglial cells). The effects of the anti-MA Fab intrabodies on the nuclear import of the preintegration complex can be more accurately studied in these cells since the nucleopores remain intact.[81-83,100,101] Similar karyophilic properties have been shown for the accessory viral protein Vpr and in future studies using anti-MA Fab intrabody, expressing nondividing cells, the relative contributions of the partly redundant MA and Vpr karyophilic localization signals could be examined in Vpr+ and Vpr− viruses.[28,84]

Antireverse Transcriptase

One study has been reported that used two selectable drug markers to establish stably transfected CD4+ T cell lines that express on separate plasmids the Fd heavy and light chains of an Fab intrabody against HIV-1 RT.[8] In these studies, inhibition of an early event of the viral life cycle, reverse transcription of viral RNA to DNA was reported. Interestingly, the anti-RT MAb that was used in these studies was not neutralizing in assays of enzyme activity. The authors proposed that the anti-RT Fab intrabodies may sterically hinder movement of the RT along the RNA template or otherwise disturb RT secondary structure. These studies did not establish how the anti-RT Fab intrabodies penetrate the tight ribonucleoprotein-RNA-tRNA-NTP complex.

Cellular Targets for Gene Therapy of HIV-1 Infection and AIDS

Future studies on the use of intrabodies to inhibit HIV-1 replication will be additionally focused on cellular target molecules that may be less prone to the somatic mutations that occur so frequently in the HIV-1 genes due to the low fidelity of HIV-1 RT. Among these target molecules will be a new group of recently discovered cell surface molecules termed the HIV-1 "coreceptors", may be excellent targets for intrabody based gene therapy approaches for HIV-1 infection and AIDS. Indeed, blocking viral entry may be the most efficient mechanism to prevent HIV-1 infection and may be favored even above blocking other intracellular viral molecules that lead to integration of the provirus.

The historical background leading to the discovery of the HIV-1 coreceptors deserves a brief introduction. Although it was well known that activated CD8+ T cells from the peripheral blood of HIV-1-infected individuals secreted one or more soluble HIV-suppressive factors (HIV-SF) that were thought to contribute to the control of HIV-1 infection in vivo,[112-115] Cocchi and colleagues[116] were the first to identify these HIV-SF as the chemokines RANTES, MIP-1a, and MIP-1β. Their findings suggested that the mechanism of action of these proinflammatory chemokines was to prevent viral entry. This observation along with the long-standing observation that the primary receptor for HIV-1, CD4, supported viral entry only when expressed on a human cell type[117-123] lead several laboratories to identify surface molecules that are now collectively known as HIV-1 coreceptors. Initially, Feng and colleagues[124] identified through a functional cDNA cloning strategy, a seven-transmembrane G protein-coupled receptor named "fusin" as a coreceptor preferentially for T cell line-tropic strains of HIV-1. Subsequently, several laboratories simultaneously discovered that CCR5, a receptor for the β-chemokines RANTES, MIP-1a and MIP-1β, served as a coreceptor for primary macrophage-tropic strains of HIV-1.[125-128] A homozygous defect in CCR5 has also been described which appears to account for resistance of some multiple-exposed individuals to HIV-1 infection.[129] Importantly, this mutation had no obvious phenotypic defect in the affected individuals. The absence of a phenotype associated with the CCR5 defect may result from the redundant nature of the chemokine system. CCR3 and CCR2B also facilitated viral entry of macrophage-tropic strains of HIV-1, but with a more restricted subset of primary viruses.[127,128]

Intrabodies are particularly well suited to cause "phenotypic knock-out" of viral and cellular proteins such as the HIV-1 coreceptors from terminally trafficking through the secretory system.[130] Several examples to support this statement have been reported.[17,131-133] In theory, there are multiple points within the secretory pathway at which an intrabody could be placed to bind and divert a trafficking protein from its ultimate destination. The ER may be the most strategic location as the tubular architecture of this organelle, combined with the precise channeling of proteins through the secretory pathway, should maximize the chances of interaction between a resident intrabody and the target protein. Unlike more distal parts of the secretory system, peptide signals required for the ER-retention of soluble proteins are well characterized.[15] The ER is also the natural site of antibody assembly being residence to molecular chaperones such as BiP and GRP94, which assist in the correct folding of immunoglobulin molecules,[134] and lastly, offers the advantage that ER-resident proteins often show extended half lives.

Other cellular targets of intrabodies may be exploited for HIV-1 gene therapy as well. For example, it is possible that G-protein mediated signal transduction events that result in membrane fusion, occur following HIV-1 binding to the G-protein coupled HIV-1 coreceptors and to CD4. If this prediction proves to be the case, then targeting these signal transducing molecules with intrabodies may block the membrane fusion event or have some negative modulating effect on this process. Likewise, HIV-1 apoptosis is probably mediated through the combined effect of viral, as well as, cellular proteins and would presumably be inhibited by directing intrabodies to these cellular proteins as well.[135] Candidate cellular proteins could include Grb3-3,[136] interleukin-1β (IL-1β)-converting enzyme (ICE), related molecules,[137,138] as well as, other cellular proteins that mediate apoptosis.[139,140]

Summary

From this review, it should be evident that intrabodies are powerful new research tools and therapeutic molecules for gene therapy of infectious diseases. Using HIV-1 as an example, intrabodies demonstrate broad versatility in their ability to inhibit different stages of the viral life cycle by targeting structural, regulatory and enzymatic proteins of the virus. Two reasons for the potent inhibitory effects of the anti-HIV-1 intrabodies may reside in the relative ease of directing the intrabodies to relevant subcellular compartments and to different epitopes on a target protein.[4] Stable expression of Sfv, Sfv-fusion protein and Fab intrabodies has been achieved in multiple subcellular compartments including, ER, distal secretory pathway, cytoplasm, nucleus and mitochondria. Even in the reducing environment of the cytoplasm where inter- and intrachain disulfide bond formation may not be optimal,[141] a sufficient number of stably expressed functional molecules are formed to inhibit HIV-1 infection.

Several potential problems exist that may limit the use of intrabodies in the clinical setting for the treatment of HIV-1 infection and AIDS. One potential problem mentioned previously is the development of HIV-1 escape mutants. Thus, for anti-HIV-1 intrabodies to have a long term effect of suppressing HIV-1 replication in the clinical gene therapy setting, epitopes on the target viral proteins will have to be carefully chosen so that if mutations do occur, the mutations are more likely to result in loss of viral protein function. Second, combinations of targets will probably have to be chosen to cripple several steps in the viral life cycle simultaneously. In future studies, bicistronic expression vectors as we have described for the expression of the Fab intrabodies, should be useful in the rapid evaluation of combination anti-HIV Sfv intrabody based gene therapy strategies. Indeed, such combination target strategies will probably be needed in the clinical gene therapy setting as recent treatment advances with combination anti-HIV-1 chemotherapy have shown.[142-144] Targeting intrabodies to cellular proteins that are involved in the HIV-1 life cycle may be a powerful additional therapeutic use of intrabodies, and may avoid the potential mutational problems that arise from targeting the viral proteins directly. Third, it remains to be determined whether the transduced cells that express intrabodies will present intrabody fragments in the context of MHC class I molecules and elicit cytotoxic T cell responses. This would be expected to be minimized with the use of human intrabodies, but will have to be determined in a human clinical trial. Finally, if long term protection of intrabody expressing, transduced $CD4^+$ and differentiated $CD34^+$ stem cells is to be achieved, these cells will have to give rise to HIV-1-resistant progeny cells which are capable of expanding in the host and allowing for long term immunologic reconstitution. Perhaps this gene therapy approach will find its place amongst the cost effective treatment and prevention strategies that are being evaluated on the global basis if the early scientific goals are set at codevelopment of in vivo injectable gene transfer vectors that will target and stably transduce HIV-1 susceptible $CD4^+$ cells and/or $CD34^+$ stem cells.

References

1. Baltimore D. Intracellular immunization. Nature 1988; 335:395-396.
2. Gilboa E, Smith C. Gene therapy for infectious diseases: the AIDS model. TIG 1994; 10:139-144.
3. Marasco WA, Haseltine WA, Chen SY. Design, intracellular expression, and activity of a human anti-human immunodeficiency virus type 1 gp120 single-chain antibody. Proc Natl Acad Sci USA 1993; 90:7889-7893.
4. Marasco WA. Intracellular antibodies (intrabodies) as research reagents and therapeutic molecules for gene therapy. Immunotechnology 1995; 1:1-19.

5. Duan L, Bagasra O, Laughlin MA et al. Potent inhibition of human immunodeficiency virus type 1 replication by an intracellular anti-Rev single-chain antibody. Proc Natl Acad Sci USA 1994; 91:5075-5079.

6. Duan L, Zhang H, Oakes JW et al. Molecular and virological effects of intracellular anti-rev single-chain variable fragments on the expression of various human immunodeficiency virus-1 strains. Human Gene Therapy 1994b; 5:1315-1324.

7. Duan L, Zhu M, Bagasra O et al. Intracellular immunization against HIV-1 infection of human T lymphocytes: Utility of anti-rev single-chain variable fragments. Hum Gene Therapy 1995; 6:1561-1573.

8. Maciejewski JP, Weichold FF, Young NS et al. Intracellular expression of antibody fragments directed against HIV reverse transcriptase prevents HIV infection in vitro. Nature Medicine 1995; 1:667-673.

9. Marasco WA 1996. Using intracellular antibodies (intrabodies) for research and gene therapy applications. In: Program and abstracts of the Gene Therapy of Cancer, AIDS and Genetic Disorders, International Symposium, International Centre for Genetic Engineering and Biotechnology, April 10-13, 1996. Trieste, Italy. (abstract).

10. Jiang W, Venugopal K, Gould EA. Intracellular interference of tick-borne flavivirus infection by using a single-chain antibody fragment delivered by recombinant Sindbis virus. J Virol 1995; 69:1044-1049.

11. Posner MR, T. Hideshima, T. Cannon, M. Mukherjee, KH Mayer KH, Byrn RA. An IgG human monoclonal antibody that reacts with HIV-1/gp120, inhibits virus binding to cells, and neutralizes infection. J Immunol 1991; 146:4325-4332.

12. Marasco WA, Bagley J, Zani C et al. Characterization of the CDNA of a broadly reactive neutralizing human anti-gp120 monoclonal antibody. J Clin Invest 1992; 90:1467-1478.

13. Thali M, Moore JP, Furman C et al. Characterization of conserved human immunodeficiency virus type 1 gp120 neutralization epitopes exposed upon gp120-CD4 binding. J Virol 1993; 67:3978-3988.

14. Chen S-Y, Bagley J, Marasco WA. Intracellular antibodies as a new class of therapeutic molecules for gene therapy. Human Gene Therapy 1994; 5:595-601.

15. Munro S, Pelham HRB A C-terminal signal prevents secretion of luminal ER proteins. Cell 1987; 48:899-907.

16. Bole DG, Hendershot LM, Kearney JF. Posttranslational association of immunoglobulin heavy chain binding protein with nascent heavy chains in nonsecreting and secreting hybridomas. J Cell Biol 1986; 102:1558-1566.

17. Richardson JH, Sodroski JG, Waldmann TA et al. Phenotypic knockout of the high-affinity human interleukin 2 receptor by intracellular single-chain antibodies against the a subunit of the receptor. Proc Natl Acad Sci USA 1995; 92:3137-3141.

18. Hoxie JA, Alpers JD, Rackowski JL et al. Alterations in T4 (CD4) protein and mRNA synthesis in cells infected with HIV. Science 1986; 234:1123-1127.

19. Buonocore L, Rose JK. Prevention of HIV-1 glycoprotein transport by soluble CD4 retained in the endoplasmic reticulum. Nature (London) 1990; 343:625-628.

20. Crise B, Buonocore L, Rose JK. CD4 is retained in the endoplasmic reticulum by the human immunodeficiency virus type 1 glycoprotein precursor. J Virol 1990; 64:5585-5593.

21. Jones SD, Porter-Brooks J, Eberhardt B et al. Gene Therapy for HIV using intracellular antibodies. J Cell Bio 1995; 21A:395.

22. Poznansky MC, Foxall R, Ramstedt U et al. Intracellular antibodies against Tat and gp120 inhibit the replication of HIV-1 in T-cells from HIV-infected individuals. Abstract in Xth International AIDS Conference (Vancouver) 1996.

23. Chen S-Y, Khouri Y, Bagley J et al. Combined intra- and extracellular immunization against human immunodeficiency virus type 1 infection with a human anti-gp120 antibody. Proc Natl Acad Sci USA 1994; 91:5932-5936.

24. Buonocore L, Rose JK. Blockade of human immunodeficiency virus type 1 production in CD4$^+$ T cells by an intracellular CD4 expressed under control of the viral long terminal repeat. Proc Natl Acad Sci USA 1993; 90:2695-2699.

25. Muzyczka, N. Use of adeno-associated virus as a general transduction vector for mammalian cells. Curr Top Microbiol Immunol 1992; 158:97-129.

26. Kotin RM. Prospects for the use of adeno-associated virus as a vector for human gene therapy. Human Gene Therapy 1994; 5:793-801.

27. Roe TY, Reynolds TC, Yu G, Brown PO. Integration of murine leukemia virus DNA depends on mitosis. EMBO J 1993; 12:2099-2108.

28. Emerman M, Bukrinsky M, Stevenson M. HIV-1 infection of nondividing cells. Nature 1994; 369:107-108.

29. Lewis P, Hensel M, Emerman M. Human immunodeficiency virus infection of cells arrested in the cell cycle. EMBO J 1992; 11:3053-3058.

30. Lewis P, Emerman M. Passage through mitosis is required for oncoretroviruses but not for the human immunodeficiency virus. J Virol 1994; 68:510-516.

31. Levin R, Mhashilkar AM, Dorfman T et al. Inhibition of early and late events of the HIV-1 replication cycle by cytoplasmic Fab intrabodies against the matrix protein, p17. 1996; submitted.

32. Elroy-Stein O, Fuerst TR, Moss B. Cap-independent translation of MRNA conferred by encephalomyocarditis virus 5' sequence improves the performance of the vaccinia virus/bacteriophage T7 hybrid expression system. Proc Natl Acad Sci USA 1989; 86:6126-6130.

33. Duke GM, Hoffman MA, Palmenberg AC. Sequence and structural elements that contribute to efficient encephalomyocarditis virus RNA translation. J Virol 1992; 66:1602-1609.

34. Chen J, Yang Q, Yang A-G et al. Intra- and extracellular immunization against HIV-1 infection with lymphocytes transduced with an AAV vector expressing a human anti-gp120 antibody gene. Human Gene Therapy 1996; (in press).

35. Malim MH, Hauber J, Fenrick R et al. Immunodeficiency virus rev trans-activator modulates the expression of the viral regulatory genes. Nature 1988; 335:181-183.

36. Laspia MF, Rice AP, Mathews MB. HIV-1 Tat protein increases transcriptional initiation and stabilizes elongation. Cell 1989; 59:282-292.

37. Marciniak RA, Garcia-Blanco MA, Sharp PA. Identification and characterization of a HeLa nuclear protein that specifically binds to the trans-activation-response (TAR) element of human immunodeficiency virus. Proc Natl Acad Sci USA 1990; 87:3624-3638.

38. Greenblatt J, Nodwell JR, Mason SW. Transcriptional anti-termination. Nature 1993; 364:401-406.

39. Madore SJ, Cullen BR. Genetic analysis of the cofactor requirement for human immunodeficiency virus type 1 Tat function. J Virol 1993; 67:3703-3711.

40. Shibuya H, Irie K, Ninomlya-Tsuji J et al. New human gene encoding a positive modulator of HIV Tat-mediated transactivation. Nature 1992; 357:700-702.

41. Ohana B, Moore PA, Ruben SR et al. The type 1 human immunodeficiency virus Tat binding protein is a transcriptional activator belonging to an additional family of evolutionarily conserved genes. Proc Natl Sci USA 1993; 90:138-142.

42. Kashanchi F, Piras G, Rodonovich MF. Direct interaction of human TFIID with the HIV-1 transactivator tat. Nature 1994; 367:295-299.

43. Gaynor R, Soultanaki E, Kuwabara M et al. Specific binding of a HeLa cell nuclear protein to RNA sequences in the human immunodeficiency virus transactivating region. Proc Natl Acad Sci USA 1989; 86:4858-4862.

44. Gatignol A, Buckler-White A, Beckhout B et al. Characterization of a human TAR RNA-binding protein that activates the HIV-1 LTR. Science 1991; 251:1597-1600.

45. Sheline CT, Milocco LH, Jones KA. Two distinct nuclear transcription factors recognize loop and bulge residues of the HIV-1 TAR RNA hairpin. Genes Dev 1991; 8:2508-2520.

46. Marciniak RA, Sharp PA. HIV-1 Tat protein promotes formation of more-processive elongation complexes. EMBO J 1991; 10:4189-4196.

47. Kato H, Sumimoto H, Pognonec P et al. HIV-1 Tat acts as a processivity factor in vitro in conjunction with cellular elongation factors. Genes Dev 1992; 6:655-666.

48. Sastry KJ, Reddy RHR, Pandita R, Totpal K, Aggarwal BB. HIV-1 Tat gene induces tumor necrosis factor-β (Lymphotoxin) in a human B-lymphoblastoid cell line. J Biol Chem 1990; 265:20091-20093.

49. Buonaguro L, Barillari G, Chang HK et al. Effects of the human immunodeficiency virus Type 1 Tat protein on the expression of inflammatory cytokines. J Virol 1992; 66:7159-7167.

50. Buonaguro L, Buonaguro FM, Giraldo G et al. The human immunodeficiency virus type 1 Tat protein transactivates tumor necrosis factor beta gene expression through a TAR-like structure. J Virol 1994; 68:2677-2682.

51. Westendorp MO, Li-Weber M, Frank RW et al. Human immunodeficiency virus type 1 Tat upregulates interleukin-2 secretion in activated T cells. J Virol 1994; 68:4177-4185.

52. Liu J, Perkins ND, Schmid RM, Nabel GJ. Specific NF-κB subunits act in concert with Tat to stimulate human immunodeficiency virus type 1 transcription. J Virol 1992; 66:3883-3887.

53. Biswas DK, Salas TR, Wang F et al. A Tat-induced auto-up-regulatory loop for superactivation of the human immunodeficiency virus type 1 promoter. J Virol 1995; 69:7437-7444.

54. Alcami J, Lera T, Folgueira L et al. Absolute dependence on NF-κB responsive elements for initiation and Tat-mediated amplification of HIV transcription in blood CD4 lymphocytes. EMBO J 1995; 14:1552-1560.

55. Biswas DK, Ahlers CM, Dezube BJ et al. Cooperative inhibition of NF-κB and Tat-induced superactivation of human immunodeficiency virus type 1 long terminal repeat. Proc Natl Acad Sci USA 1993; 90:11044-11048.

56. Kamine J, Chinnadurai G. Synergistic activation of the human immunodeficiency virus type 1 promoter by the viral Tat protein and cellular transcription factor Sp1. J Virol 1992; 66:3932-3936.

57. Dayton AI, Sodroski JG, Rosen CA et al. The trans-activator gene of the human T cell lymphocytic virus type III is required for replication. Cell 1986; 44:941-947.

58. Fisher AG, Feinberg MB, Joseph SF. The trans-activator gene of HTLV-III is essential for virus replication. Nature (London) 1986; 320:361-371.

59. Mhashilkar AM, Bagley J, Chen SY et al. Inhibition of HIV-1 Tat-mediated LTR transactivation and HIV-1 infection by anti-Tat single chain intrabodies. EMBO J 1995; 14:1542-1551.

60. Kozak M. Point mutations define a sequence flanking the AUG initiator codon that modulates translation by eukaryotic ribosomes. Cell 1986; 44:283-292.

61. Pepys MB. In: MM Frank, KF Austen, HN Clamen et al, eds. "Amyloidosis". Samter's Immunologic Diseases. Vol. 1. 5th edition. Boston: Little, Brown & Co., 1995:637-656.

62. Biocca S, Neuberger MS, Cattaneo A Expression and targeting of intracellular antibodies in mammalian cells. EMBO J 1990; 1:101-108.

63. Biswas DK, Dezube BJ, Ahlers CM et al. Pentoxifylline inhibits HIV-1 LTR-driven gene expression by blocking NF-κB action. J AIDS 1993; 6:778-786.

64. Biswas DK, Ahlers CM, Dezube BJ et al. Pentoxifylline and other protein kinase C inhibitors down-regulate HIV-LTR NF-κB induced gene expression. Mol Med 1994; 1:31-43.

65. Qatsha KA, Rudolph C, Marmé D et al. Gö 6976, a selective inhibitor of protein kinase C, is a potent antagonist of human immunodeficiency virus 1 induction from latent/low-level-producing reservoir cells in vitro. Proc Natl Acad Sci USA 1993; 90:4674-4678.

66. Haseltine WA Molecular biology of the human immunodeficiency virus type 1. FASEB J 1991; 5:2349-2360.

67. Olsen HS, Cochrane AW, Dillon PJ et al. Interaction of the human immunodeficiency virus type 1 Rev protein with a structured region in *env* MRNA is dependent on multimer formation mediated through a basic stretch of amino acids. Genes & Dev 1990; 4:1357-1364.

68. Malim MH, Cullen BR. HIV-1 structural gene expression requires the binding of multiple Rev monomers to the viral RRE: implications for HIV-1 latency. Cell 1991; 65:241-248.

69. Frankhauser C, Izaurralde E, Adachi Y et al. Specific complex of human immunodeficiency virus type 1 Rev and Nucleolar B23 proteins: dissociation by the Rev response element. Mol Cell Biol 1991; 11:2567-2575.

70. Fritz CC, Zapp ML, Green MR. A human nucleoporin-like protein that specifically interacts with HIV Rev. Nature 1995; 376:530-533.

71. Kalland K-H, Szilvay AM, Brokstad KA et al. The human immunodeficiency virus type 1 Rev protein shuttles between the cytoplasm and nuclear compartments. Mol Cell Biol 1994; 14:7436-7444.

72. Kalland K-H, Szilvay AM, E Langhoff et al. Subcellular distribution of human immunodeficiency virus type 1 Rev and colocalization of Rev with RNA splicing factors in a speckled pattern in the nucleoplasm. J Virol 1994; 68:1475-1485.

73. Marasco WA, Szilvay AM, Kalland KH et al. Spatial association of HIV-1 tat protein and the nucleolar transport protein B23 in stably transfected Jurkat T-cells. Arch Virol 1994; 139:133-154.

74. Meyer BB, Malim MH. The HIV-1 Rev *trans*-activator shuttles between the nucleus and the cytoplasm. Genes & Dev 1994; 8:1538-1547.

75. Ruhl M, Himmelspach M, Bahr GM et al. Eukaryotic initiation factor 5A is a cellular target of the human immunodeficiency virus type 1 Rev activation domain mediating *trans*-activation. J Cell Biol 1993; 123:1309-1320.

76. Luo Y, Yu H, Peterlin BM Cellular protein modulates effects of human immunodeficiency virus type 1 rev. J Virol 1994; 68:3850-3856.

77. Shukla RR, Kimmel PL, Kumar A. Human immunodeficiency virus type 1 Rev-responsive element RNA binds to host cell-specific proteins. J Virol 1994; 68:2224-2229.

78. Park H, Davies MV, Langland JO et al. TAR RNA-binding protein is an inhibitor of the interferon-induced protein kinase PKR. Proc Natl Acad Sci USA 1994; 91:4713-4717.

79. Wu Y, Duan L, Zhu M et al. Binding of intracellular anti-Rev single chain variable fragments to different epitopes of human immunodeficiency virus type 1 Rev: variations in viral inhibition. J Virol 1996; 70:3290-3297.

80. Bukrinskaya AG, Vorkunova GK, Tentsov YY. HIV-1 matrix protein P17 resides in cell nuclei in association with genomic RNA. AIDS Res And Hum Retroviruses 1992; 8:1795-1801.

81. Bukrinsky MI, Sharova N, McDonald TL et al. Association of integrase, matrix, and reverse transcriptase antigens of human immunodeficiency virus type 1 with viral nucleic acids following acute infection. Proc Natl Acad Sci USA 1993a; 90:6125-6129.

82. Bukrinsky MI, Sharova N, Dempsey M et al. Active nuclear import of human immuno-deficiency virus type 1 preintegration complexes. Proc Natl Acad Sci USA 1992; 89:6580-6584.

83. von Schwedler U, Kornbluth RS, Trono D. The nuclear localization signal of the matrix protein of human immunodeficiency virus type 1 allows the establishment of infection in macrophages and quiescent T lymphocytes. Proc Natl Acad Sci USA 1994; 91:6992-6996.

84. Heinzinger NK, Bukrinsky MI, Haggerty SA et al. The Vpr protein of human immunodeficiency virus type 1 influences nuclear localization of viral nucleid acids in nondividing host cells. Proc Natl Acad Sci USA 1994; 91:7311-7315.

85. Reicin AS, Paik S, Berkowitz RD et al. Linker insertion mutations in the human immunodeficiency virus type 1 *gag* gene: Effects on virion particle assembly, release, and infectivity. J Virol 1995; 69:642-650.

86. Aldovini A, Young RA. Mutations of RNA and protein sequences involved in human immunodeficiency virus type 1 packaging result in production of noninfectious virus. J Virol 1990; 64:1920-1926.

87. Gorelick RJ, Nigida Jr SM, Bess Jr JW. Noninfectious human immunodeficiency virus type 1 mutants deficient in genomic RNA. J Virol 1990; 64:3207-3211.

88. Rice WG, Supko JG, Malspeis L et al. Inhibitors of HIV nucleocapsid protein zinc fingers as candidates for the treatment of AIDS. Science 1995; 270:1194-1197.

89. Li X, Quan Y, Arts EJ et al. Human immunodeficiency virus type 1 nucleocapsid protein (Ncp7) directs specific initiation of minus-strand DNA synthesis primed by human TRNA$_3^{Lys}$ in vitro: Studies of viral RNA molecules mutated in regions that flank the primer binding site. J Virol 1996; 70:4996-5004.

90. Sharova N, Bukrinskaya A. P17 and p17-containing *gag* precursors of input human immunodeficiency virus are transported into the nuclei of infected cells. AIDS Res Hum Retrovir 1991; 7:303-306.

91. Spearman P, Wang J-J, Vander Heyden N et al. Identification of human immunodeficiency virus type 1 gag protein domains essential to membrane binding and particle assembly. J Virol 1994; 68:3232-3242.

92. Zhou W, Parent LJ, Wills JW et al. Identification of a membrane-binding domain within the amino terminal region of human immunodeficiency virus type 1 gag protein which interacts with acidic phospholipids. J Virol 1994; 68:2556-2569.

93. Göttlinger HG, Sodroski JG, Haseltine WA. Role of capsid precursor processing and myristolation in morphogenesis and infectivity of human immunodeficiency virus type 1. Proc Natl Acad Sci USA 1989; 86:5781-5785.

94. Bryant M, Ratner L. Myristoylation-dependent replication and assembly of human immunodeficiency virus 1. Proc Natl Acad Sci USA 1990; 87:523-527.

95. Pal R, Reitz Jr MS, Tschachler RC et al. Myristoylation of *gag* proteins of HIV-1 plays an important role in virus assembly. AIDS Res Hum Retrovir 1990; 6:721-730.

96. Dorfman T, Mammano F, Haseltine WA et al. Role of the matrix protein in the virion association of the human immunodeficiency virus type 1 envelope glycoprotein. J Virol 1994; 68:1689-1696.

97. Yu X, Yuan X, Matsuda Z et al. The matrix protein of human immunodeficiency virus type 1 is required for incorporation of viral envelope protein into mature virions. J Virol 1992; 66:4966-4971.

98. Yu X, Yu QC, Lee TH et al. The C terminus of human immunodeficiency virus type 1 matrix protein is involved in early steps of virus life cycle. J Virol 1992; 66:5667-5670.

99. Bukrinskaya AG, Ghorpade A, Heinzinger NK et al. Phosphorylation-dependent human immunodeficiency virus type 1 infection and nuclear targeting of viral DNA. Proc Natl Acad Sci USA 1996; 93:367-371.

100. Gallay P, Swingler S, Aiken C. HIV-1 infection of nondividing cells: C-terminal tyrosine phosphorylation of the viral matrix protein is a key regulator. Cell 1995; 80:379-388.

101. Gallay P, Swingler S, Song J et al. HIV nuclear import is governed by the phosphotyrosine-mediated binding of matrix to the core domain of integrase. Cell 1995; 83:569-576.

102. Niedrig M, Hinkula J, Weigelt W et al. Epitope mapping of monoclonal antibodies against human immunodeficiency virus type 1 structural proteins by using peptides. J Virol 1989; 63:3525-3528.

103. Louwagie J, McCutchan FE, Peeters M et al. Phylogenetic analysis of gag genes from 70 international HIV-1 isolates provides evidence for multiple genotypes. AIDS 1993; 7:769-780.

104. Helseth EM, Kowalski D, Gabuzda U. Rapid complementation assays measuring replicative potential of human immunodeficiency virus type 1 envelope glycoprotein mutants. J Virol 1990; 64:2416-2420.

105. Ansari-Lari MA, Donehower LA, Gibbs RA. Analysis of human immunodeficiency virus type 1 integrase mutants. Virology 1995; 211:332-335.

106. Wiskerchen M, Muesing MA Human immunodeficiency virus type 1 integrase: effects of mutations on viral ability to integrate, direct viral gene expression rom unintegrated viral DNA templates, and sustain viral propagation in primary cells. J Virol 1995; 69:376-386.

107. Sakai H, Kawamura M, Sakuragi JI et al. Integration is essential for efficient gene expression of human immunodeficiency virus type 1. J Virol 1993; 67:1169-1174.

108. Engelman A, Englund G, Orenstein JM et al. Multiple effects of mutations in human immunodeficiency virus type 1 integrase on viral replication. J Virol 1995; 69:2729-2736.

109. Stevenson M, Haggerty S, Lamonica CA et al. Integration is not necessary for expression of human immunodeficiency virus type 1 protein products. J Virol 1990; 64:2421-2425.

110. Myers G, Korber B, Wain-Hobson S et al. Human retroviruses and AIDS: a compilation and analysis of nucleic acid and amino acid sequences. Los Alamos National Laboratory, N.M. 1994.

111. Freed E, Orenstein JM, Buckler-White JM. Single amino acid changes in the human immunodeficiency virus type 1 matrix protein block virus particle production. J Virol 1994:5311-5320.

112. Walker CM, Moody DJ, Stites DP, Levy JA. CD8+ lymphocytes can control HIV infection in vitro by suppressing virus replication. Science 1986; 234:1563-1566.

113. Brinchmann JE, Gaudernack G, Vartdal F. CD8+ cells inhibit HIV replication in naturally infected CD4+ T cells. J Immunol 1990; 144:2961.

114. Mackewicz CE, Levy JA. CD8+ cell anti-HIV activity: Nonlytic suppression of virus replication. AIDS Res Hum Retroviruses 1992; 8:1039.

115. Mackewicz CE, Ortega HW, Levy JA. CD8$^+$ cell anti-HIV activity correlates with the clinical state of the infected individual. J Clin Invest 1991; 87:1462-1466.

116. Cocchi F, DeVico AL, Garzino-Demo A. Identification of RANTES, MIP-1a, and MIP-1β as the major HIV-suppressive factors produced by CD8$^+$ T Cells. Science 1995; 270:1811-1815.

117. Weiner DB, H. Huebner, Williams WV, Greene MI. Human genes other than CD4 facilitate HIV-1 infection of murine cells. Pathobiology 1991; 59:361-371.

118. Dragic T, Charneau P, Clavel F et al. Complementation of murine cells for human immunodeficiency virus envelope/CD4-mediated fusion in human/murine heterokaryons. J Virol 1992; 66:4794-4802.

119. Broder CC, Dimitrov DS, Blumenthal R et al. The block to HIV-1 envelope glycoprotein-mediated membrane fusion in animal cells expressing human CD4 can be overcome by a human cell component(s). Virology 1993; 193:483-491.

120. Harrington RD, Geballe AP. Cofactor requirement for human immunodeficiency virus type 1 entry into a CD4-expressing human cell line. J Virol 1993; 67:5939-5947.

121. Ramarli D, Cambiaggi C, De Giuli CM, Tripputi P. Susceptibility of human-mouse T cell hybrids to HIV-productive infection. AIDS Res Hum Retroviruses 1993; 9:1269-1275.

122. Dragic T, Picard L, Alizon M. Proteinase-resistant factors in human erythrocyte membranes mediate CD4-dependent fusion with cells expressing human immunodeficiency virus type 1 envelope glycoproteins. J Virol 1995; 69:1013-1018.

123. Dragic T, Alizon M. Different requirements for membrane fusion mediated by the envelopes of human immunodeficiency virus types 1 and 2. J Virol 1993; 67:2355-2359.

124. Feng Y, Broder CC, Kennedy PE. HIV-1 entry cofactor: functional CDNA cloning of a seven-transmembrane, G protein-coupled receptor. Science 1996; 272:872-877.

125. Deng HK, Liu R, Ellmeier W et al. Identification of a major coreceptor for primary isolates of HIV-1. Nature 1996; 381:661-666.

126. Dragic T, Litwin V, Allaway GP et al. HIV-1 entry into CD4$^+$ cells is mediated by the chemokine receptor CC-CKR-5. Nature 1996; 381:667-673.

127. Choe H, Farzan M, Sun Y et al. The β-chemokine receptors CCR3 and CCR5 facilitate infection by primary HIV-1 isolates. Cell 1996; 85:1135-1148.

128. Doranz BJ, Rucker J, Yi Y et al. A dual-tropic primary HIV-1 isolate that uses fusin and the β-chemokine receptors CKR-5, CKR-3, and CKR-2b as fusion cofactors. Cell 1996; 85:1149-1158.

129. Liu R, Paxton WA, Choe S,.Ceradini D et al. Homozygous defect in HIV-1 coreceptor accounts for resistance of some multiply-exposed individuals to HIV-1 infection. Cell 1996; 86:367-377.

130. Richardson JH, Marasco WA. Intracellular antibodies: development and therapeutic potential. TIBTech 1995; 13:306-310.

131. Deshane J, Loechel F, Conry RM et al. Intracellular single-chain antibody directed against erbB2 down-regulates cell surface erbB2 and exhibits 1 selective anti-proliferative effect in erbB2 overexpressing cancer cell lines. Gene Therapy 1994; 1:332-337.

132. Beerli RR, Wels W, Hynes NE. Intracellular expression of single chain antibodies reverts ErbB-2 transformation. J Biol Chem 1994; 269:23931-23936.

133. Graus-Porta D, Beerli RR, Hynes NE. Single-chain antibody-mediated intracellular retention of ErbB-2 impairs neu differentiation factor and epidermal growth factor signaling. Mol Cell Biol 1995; 15:1182-1191.

134. Melnick J, Aviel S, Argon Y. The endoplasmic reticulum stress protein GRP94, in addition to BiP, associates with unassembled immunoglobulin chains. J Biol Chem 1992; 267:21303-21306.

135. Pantaleo G, Fauci AS. Apoptosis in HIV infection. Nature Medicine 1995; 1:118-120.
136. Fath I, Schweighoffer F, Rey I et al. Cloning of a Grb2 isoform with apoptotic properties. Science 1994; 264:971-974.
137. Muzio M, Chinnaiyan AM, Kischkel FC et al. FLICE, a novel FADD-homologous ICE/CED-3-like protease, is recruited to the CD95 (Fas/APO-1) death-inducing signaling complex. Cell 1996; 85:817-827.
138. Boldin MP, Goncharov TM, Goltsev YV et al. Involvement of MACH, a novel MORT1/FADD-interacting protease, in Fas/APO-1- and TNF receptor-induced cell death. Cell 1996; 85:803-815.
139. Barr PJ, Tomei LD. Apoptosis and its role in human disease. Bio/Technology 1994; 12:487-493.
140. Thompson CB. Apoptosis in the pathogenesis and treatment of disease. Science 1995; 267:1456-1462.
141. Biocca S, Ruberti F, Tafani M et al. Redox state of single chain Fv fragments targeted to the endoplasmic reticulum, cytosol and mitochondria. Biotechnology 1995; 13:1110-1115.
142. Choo V. Combination superior to zidovudine in Delta trial. Lancet 1995 346:895.
143. Collier AC, Coombs RW, Schoenfeld DA et al. Extended treatment with saquinavir (SAQ), zidovudine (ZDV), and zalcitabine (ddc) vs SAQ and ZDV vs ddc and ZDV. In: Program and abstracts of the 35th Interscience conference on Antimicrobial Agents and Chemotherapy, San Francisco, September 17-20, 1995. Washington, D.C.: American Society for Microbiology, 236 (abstract).
144. Collier AC, Coombs RW, Schoenfeld DA et al. Treatment of human immunodeficiency virus infection with saquinavir, zidovudine, and zalcitabine. NEJ Med 1996; 334:1011-1017.
145. Mhashilkar AM, Biswas DK, Pardee AB, Marasco WA. Inhibition of HIV-1 replication in vitro by a novel combination of anti-tat single chain intrabodies and NF-κB antagonists. 1996; submitted.

CHAPTER 9

Plantibodies: Immunomodulation and Immunotherapeutic Potential

Rosella Franconi, Paraskevi Tavladoraki and Eugenio Benvenuto

Plants are currently being explored and exploited as an important system for the expression of recombinant proteins from different sources and represent the next wave in the production of bioactive proteins, either to improve plant performance itself or to be used as biofactories of high-value therapeutic products.[1] This chapter describes recent trends and achievements in the field of "plantibodies" (plant produced antibodies), a rapidly evolving field since the original description of antibodies produced in plants.[2] Antibody engineering has proved to be a powerful means of modifying immunoglobulin genes to be functionally expressed also in biological systems different from lymphoid cells. Among these systems higher plants may well be considered effective hosts for heterologous expression. In fact, the general capability and adaptability of plants to transformation techniques and expression of immunoglobulin genes as well as to antibody processing and assembly (matching quite closely that of native cells) make plants particularly advantageous for many biotechnological applications.

Antibody production in plants is an intensive research area not only restricted to classical immunotherapy goals. We will attempt to extend the prevalent medical definition of immunotherapy giving examples of the exploitation of the vast repertoire of the vertebrate immune system to endow plants with new forms of disease defense. In this way, among other previously described applications, the expression of recombinant antibodies provides the basis of a revolutionary 'genetic immunization' against plant diseases. The interesting applications and the potential of plantibody mediated immunotherapy for plants and animals will be described.

Plant Transformation Systems

As a result of the progress made in the eighties, the vectors based on the natural gene transfer capability of *Agrobacterium tumefaciens* are now the most appropriate to stably integrate foreign genes into plants (for a review see ref. 3). In nature, the soil borne bacterium *A. tumefaciens* infects mainly dicotyledonous plants, inducing a neoplastic transformation called crown gall. A common feature of these transformed cells is the production of amino acid derivatives called opines, represent a carbon and nitrogen source for the bacterium. The genes involved in this neoplastic transformation are all contained in a bacterial sequence portion (approx. 20 Kb) called T-DNA, located on the tumor-inducing (Ti) plasmid. The T-DNA is

Intracellular Antibodies: Development and Applications, edited by
Antonino Cattaneo and Silvia Biocca. © Springer-Verlag and Landes Bioscience 1997.

transferred and integrated into plant chromosomes. It has been demonstrated that *A. tumefaciens* never enters the plant cell but responds to chemical signals produced after plant cell wounding, activating a complex system of gene transfer similar to bacterial conjugation. Two components of the Ti plasmid are essential for the transfer and integration of T-DNA: the border regions of the T-DNA and the virulence *(vir)* gene cluster, which is the main switch of the transformation process. The *vir* region (approx. 35 kb), which is located separately from the T-DNA region on the Ti plasmid, provides most of the functions for T-DNA transfer in trans. Essentially, this implies that any kind of plasmid construct containing these two functional elements (border and *vir* genes) could be used to direct the expression of foreign genes in plants. Vectors are available on which genes involved in tumorigenesis are completely removed and substituted by a cloning region usually flanked by a selectable marker gene. These vectors equipped with appropriate origins of replication and T-DNA border regions are engineered and propagated in *Escherichie coli*. Subsequently, they are transferred to an *Agrobacterium* strain containing a resident Ti plasmid devoid of oncogenic functions but bearing an intact *vir* region. Foreign genes are thus mobilized into plant cells that are eventually regenerated into a fully functional transgenic plant.

Antibody Engineering in Plants: A Chronological Survey

The pioneering work of immunoglobulin expression in plants was carried out by Andrew Hiatt and co-workers.[2] Genes from a monoclonal antibody directed against a low molecular weight phosphonate ester were cloned from mouse hybridoma cells and transgenic plant lines were obtained which expressed separately either the light (κ) or heavy (γ) chain of the IgG1 molecule. Transformed plants expressing individual chains, which carried the wild type signal sequence of the mouse immunoglobulin, were crossed to produce filial recombinants that expressed assembled functional γ-κ complexes. The authors demonstrated that signal sequences were fundamental for the production of fully functional complete immunoglobulin molecules. A different strategy was later employed to achieve coordinated expression of light and heavy chains.[4] A unique expression vector was constructed linking both genes under the control of two different plant promoters, integrating the foreign genes into the plant genome through a single transfer event mediated by *Agrobacterium*. By this procedure, sexual crosses between transgenic plants were eliminated and the simultaneous expression of light (λ) and heavy (μ) chains of the B1-8 antibody (IgM) was obtained. In this case, the antibody chains were equipped with plant signal sequences.

While efforts were being made to express whole immunoglobulins in plants, attempts were also made to overcome the requirement for the assembly of two immunoglobulin chains, designing simpler forms of antibody molecule able to immunomodulate selected functions in vivo in the relevant cellular compartment. In fact, the Fc region effector functions likely to be inactive in plants could be considered dispensable or undesirable in many plant applications. Since immunoglobulin constant regions could be removed from variable domains generally without any apparent detrimental effect on binding,[5] expression of engineered antibody forms such as single domain antibodies (VH fragment alone, dAb) or single chain antibodies (VH and VL joined by a linker peptide, scFv) have been obtained in plants.[6,7] Efficient expression of dAbs is particularly attractive due to their easy manipulation and molecular size that would facilitate access to a target protein. A serious drawback of these minimal antigen recognition units is the exposed hydrophobic surface of the molecule[8] that may require extensive engineering to avoid insolubility and nonspecific binding. In this regard, the recent discovery in camelid species[9] of a new class of antibodies naturally lacking light chains provides precious leads to the identification of motifs or residues[10,11] that might be a key to the successful utilization of functional single VH domains with desired affinities and specificities. At present, however, the most versatile antibody derivative for regulated expression in different plant tissues or cellular compartments is the single chain antibody. Thus, cytosolic expression of ScFv fragments has been successfully employed for the induction of aberrant phytochrome-dependent germination[7] and protection against virus infection.[12] Hence, this molecule seems particularly appropriate when molecular interference must be achieved in cellular compartments such as the cytoplasm of plant cells where leaderless heavy and light chains fail to assemble.[2,13]

Adjustments were made to achieve coordinate expression and higher yields of whole antibody[14] and Fab fragments[15] in different plant organs. Meanwhile, different scFvs were directed to the endoplasmic reticulum (ER) utilizing plant intracellular-trafficking signals.[16-18] It was generally found that passage into the ER can circumvent problems of instability leading to an enhanced level of protein accumulation and so far this holds true for both whole and antibody domains. In plants, unlike in animal cells, assembly of full length antibodies in the cytoplasm could not be demonstrated while ScFvs could be very efficiently expressed in the cytoplasm. Therefore, it appears clear that engineered antibodies would be more valuable for intracellular immunomodulation to remodel or immunize the plants themselves. On the other hand, secreted antibodies are very qualified for large scale production. The remarkable finding that a secretory IgA antibody, preventing streptococcal colonization in humans can be correctly assembled in transgenic tobacco cells,[19,20] while two different cell types are required in mammals (see below) opens the field of transgenic plant technology to passive immunization strategy for biomedical applications.

Antibody Targeting

Successful application of antibody expression in plants relies on the targeting of the recombinant immunoglobulin molecule of choice to a particular subcellular compartment where it should maintain its functionality, be expressed at optimized levels and interact with the antigen (in case immunomodulation is required). This is a research area where a great effort has currently been invested as the requirements for protein sorting and compartmentalization in plants, similarly to other eukaryotic organisms, are far from being completely understood (see chapter 5). At present, plantibody targeting has only been attempted to the secretory pathway and the cytoplasm, although targeting of other proteins to different compartments have been described in plants (for a review see ref. 21). Here we present a brief and certainly not exhaustive description of the plant protein secretory and sorting pathways which have been extensively covered elsewhere (for a review see refs. 21-24).

Protein Sorting in Plants

Most proteins present in the various subcellular compartments (with the exception of a few proteins made within mitochondria and chloroplasts) are synthesized on cytosolic or membrane-bound ribosomes and the transport and sorting of proteins to their ultimate site depends on specific targeting signals present on the primary sequence or the structure of the protein. The endomembrane system of plant cells consists in a series of organelles that include the ER, Golgi apparatus, vacuole and transport vesicles that mediate connections among these organelles and the plasma membrane by a mechanism of continuous budding and docking.

The plant secretory pathway has unique features not seen in yeast or animal cells. These include the capacity to communicate between cells through ER-containing plasmodesmata[25] and the potential to localize, assemble and store large protein bodies in either the ER lumen[24] or vacuole[26] (e.g., prolamines and globulins) as a part of a programmed developmental process (for a review see ref. 27). Protein body formation in plants is carried out by a series of regular events of protein synthesis and assembly of correctly folded proteins: this represents a difference to the animal system where the formation of intracisternal granules occurs as a result of the aggregation of misfolded proteins. Differences may also exist in the secretory pathway activity between various plant tissues, in particular, roots and leaves. Plant cells also secrete many soluble proteins which, may cross the plant cell wall and reach the intercellular spaces (apoplast).

The first step in the transport route common to secretory, vacuolar, ER or Golgi resident proteins is a signal-peptide mediated translocation across the ER membrane.[28] A high degree of conservation in the ER targeting machinery between the animal and plant kingdom has been described, although plant and mammalian components are not always fully interchangeable.[29,30] Signal peptides vary considerably in length and sequence between different organisms. Nevertheless, signal sequences of various origins can drive and redirect heterologous proteins in the ER of plant cells[31-34] (Tables 9.1 and 9.2). Once in the lumen of the ER, the new polypeptide may undergo ER-specific cotranslational and post-translational modification (i.e., cleavage of the signal peptide, disulfide bond formation, N-linked glycosylation). Further information for specific transport or retention in the ER are present in the domains of the protein that have undergone such modifications.[35] Such targeting information appears not to be required for secretion and a soluble protein containing only a signal peptide is transported extracellularly. Secretion is considered to be a "default pathway", as demonstrated in plant cells by Denecke et al,[32] who established an in vivo model system for protein transport based on the transient synthesis of heterologous proteins in tobacco protoplasts. However, things could be more complex as in mammalian cells (see chapter 5).

Retention in the Endoplasmic Reticulum

Also in plants, soluble ER resident proteins (collectively referred as to as reticuloplasmins) possess a carboxy-terminal sorting domain consisting of the four amino acids KDEL or a small set of related tetrapeptide variants necessary for retention in the ER lumen,[36] probably by continuous retrieval from a post-ER compartment.[37,38] Recently, an *Arabidopsis thaliona* gene which encodes a putative receptor for the KDEL-like signals, with seven trans-membrane domains and homolog to the yeast Erd 2 gene, has been cloned.[39] The mechanism for retention/retrieval seems to be conserved between plant, mammalian and yeast cells, and the KDEL-like sequences when fused to normally secreted heterologous proteins retained in the ER.[40,24] There are, however, some examples of engineered plant proteins which, although tagged with the KDEL-like sequences, are not localized in the ER (i.e., the vacuolar storage protein, phytoemoagglutinin.[41] Moreover, the KDEL sequence of the plant vacuolar protein cysteine endopeptidase (SH-EP) is cleaved by vacuolar proteases.[42]

In plants, a few number of resident integral membrane proteins have been identified. Some of them are involved in vesicular trafficking[43,44] while others are involved in solute transport across membranes (for a review see ref. 45). Little is known about the mechanisms by which membrane proteins are correctly localized and retained in the secretory pathway. Moreover, while targeting signals of several soluble proteins of the plant secretory pathway have been identified, almost nothing is known about the targeting signals within the membrane proteins.[46] It is not known whether the 'default' compartment for membrane proteins is the plasma membrane or the vacuole.

Molecular chaperones are required for the folding of newly translocated proteins but also for the translocation process itself: they prevent incorrect interactions within and between non-native polypeptides and increase the yield of correctly folded proteins.[47] In the ER lumen, plants possess several of the chaperones found in other organisms. ER resident chaperons, homologous to immunoglobulin heavy chain binding protein (BiP)[48] and HSP90 (stress-90 proteins, like Grp94),[49] have been described in plants, as well as calnexin[50] and calreticulin[51] (for a review see ref. 52). BiP in tobacco is encoded by a multigene family, is abundantly expressed in tissues with high numbers of dividing cells and is induced when plant cells are treated with the inhibitor of glycosylation tunicamycin, or are subjected to stress.[53-55] BiP also acts as a quality control monitor in binding and retaining assembly-defective,[56] malfolded[54] or unglycosylated[55] proteins in an ATP-dependent manner.[57] Studies on developing rice endosperm cells have shown that BiP facilitates translocation of the nascent chain across the ER membrane.[58] Moreover, the presence in plants of "folding catalysts" (or foldases) like disulfide isomerase (PDI)[59,60] helps to accelerate the intrinsic slow steps in the folding of some proteins, such as the formation of disulfide bonds in secretory proteins.[61]

The Golgi apparatus in plants, in contrast to the clustered Golgi stacks of animal cells, is dispersed singly or in small clusters throughout the cytoplasm. Little is known about the plant Golgi apparatus and the trans-Golgi network (TGN) at molecular level and only few enzymes and transport proteins have been purified to homogeneity (for a review see ref. 62). Protein modification by the addition of carbohydrates plays a role in protein folding and quality control (for a review see ref. 63). N-linked as well as O-linked glycosylation are present in plant proteins. N-linked glycosylation starts in the ER and terminates in the Golgi, while O-linked glycosylation is peculiar to Golgi. While no structural difference is observed among eukaryotic organisms for high mannose-type glycans, complex glycans that are generated subsequently in the Golgi may vary quite a lot from animals to plants. In fact, complex glycan tends to be smaller in plants and the addition of sialic acid, terminal residue of most mammal complex glycans, is lacking. The absence in plants of sialic acid and the corresponding sialyl transferase activity[64] constitutes a point of divergence between mammalian and plant systems.

Vacuolar Targeting Signals

Little is known about Golgi-to-plasma membrane transport in plants and it is also not clear at which point sorting to the vacuole takes place within the secretory pathway. The vacuole (comparable to the lysosome of animal cells and the vacuole

Table 9.1. Expression of full length antibodies in plants

Antibody	Construction	Signal Sequence	Species	Localization	Expression Yield % of total soluble protein	Ref.
mAb 6D4 catalytic antibody to phosphonate ester (IgG1)	F.[CaMV 35S-V_H-Cγ × CaMV 35S-V_L-C_L]	none none	*Nicotiana tabacum*	N.D.	max 0.002 H chain max 0.006 L chain (0% assembly)	2
	F.[CaMV 35S-ss-V_H-Cγ × CaMV 35S-ss-V_L-C_L]	original mouse original mouse			max 1.3 each chain (95% assembly)	
mAb B1-8 anti-NP hapten (IgM)	pNOS-ss-V_H-C_μ + pT_{R1}'-ss-V_L-C_L (Mounted on the same T-DNA)	α-amylase from barley aleurone	*N. tabacum*	Endoplasmic reticulum and chloroplast	N.D.	4
mAb 6D4 catalytic antibody to phosphonate ester (IgG1)	F.[CaMV 35S-ss-V_H-Cγ × CaMV 35S-ss-V_L-C_L]	yeast α-mating type factor (signal sequence and prosequence)	*N. tabacum*	Protoplast culture medium and cell suspensions	max 0.8	13
mAb MAK 33 anti-human creatine kinase (IgG1)	CaMV 35S-ss-V_H-C_H + [CaMV 35S-ss-V_L-C_L or CaMV 35S-ss-V_H-$C_{\gamma 1}$] (IgG and Fab fragment) (double transformation)	2S2 (*A. thaliana* storage protein)	*N. tabacum* and *Arabidopsis thaliana*	N.D. (Fab fragment detected only in the nucleolus in tobacco meristematic shoot tissue)	0.002 – 1.3 (variations between Fab and IgG as well as between *Arabidopsis* and *Nicotiana*)	15

mAb Guy's13 anti-adhesin (SA I/II) of *Streptococcus mutans* (IgG1) and further modifications to chimeric IgG/A	F_1[CaMV 35S-ss-V_H-Cγ1,2,3 or CaMV 35S-ss-V_H-Cγ1-Cα2,3 or CaMV 35S-ss-V_H-Cγ1,2-Cα2,3 x CaMV 35S-ss-V_L-C_L]	original mouse	*N. tabacum*	secretory pathway	mean 4.5 µg/ml of plant extracts	19
mAb 21C5 anti-cutinase of *Botrytis cinerea* (IgG1)	CaMV 35S (double enhancer)-ss-V_H-C_H + pT_R2'-1'-ss-V_L-C_L (pT_R2' mounted on the same T-DNA in two orientations)	mouse (slightly modified from κ-chain of mAb CEA66-E3)	*N. tabacum*	Protoplast culture medium and cell suspensions	max 0.6 (constructs in opposite orientation) max 1.1 (constructs in the same orientation)	14
mAb 24 anti-TMV coat protein (IgG2b)	CaMV 35S-ss-V_H-C_γ + CaMV 35S-ss-V_L-C_L (Mounted on the same T-DNA)	original mouse	*N. tabacum*	intercellular spaces	max 0.16 (10% of total protein from intercellular space)	83
mAb Guy's13 anti-adhesin (SA I/II) of *Streptococcus mutans* (IgG1) and further modifications to chimeric IgG/A	F_1[CaMV 35S-ss-V_H-Cγ1-Cα2,3 x CaMV 35S-ss-V_L-C_L] x CaMV 35S-ss-J x CaMV 35S-ss-SC (dimeric secretory IgA-G)	original mouse	*N. tabacum*	Mesophyll cells (mainly bundle sheath cells) not extracellular	200-500 µg/g fresh weight	20

continued...

Table 9.1. Expression of full length antibodies in plants (continued)

Antibody	Construction	Signal Sequence	Species	Localization	Expression Yield % of total soluble protein	Ref.
mAb MAK 33 anti-human creatine kinase (IgG1)	CaMV 35S-ss-V_H-C_H + [CaMV 35S-ss-V_L-C_L or CaMV 35S-ss-V_l-$C_{\gamma 1}$] (IgG and Fab fragment) (double transformation)	2S2 (*A. thaliana* storage protein)	*N. tabacum*	intercellular spaces Fab also in xylem vessels	0.2-0.4 (11-13% of total protein from intercellular space)	86
mAb 6D4 anti-salivary secretion of root-knot nematode *Meloidogyne incognita* (IgM)	F_1[CaMV 35S-ss-V_H-C_H x CaMV 35S-ss-V_L-C_L]	original mouse	*N. tabacum*	stem, leaves, roots, flowers, developing seed pods (secretion not determined)	max 0.01	120

N.D.: not determined
x: sexual crossing

Table 9.2. *Expression of antibody fragments in plants*

Antibody	Construction (scFv / dAb)	Signal Sequence	Species	Localization	Expression Yield % of total soluble protein	Ref.
mAb AS32 anti-phytochrome (IgG1)	CaMV 35S-V_L-L-V_H L: EGKSSGSGSESKP	none	*Nicotiana tabacum*	intracellular	0.06 - 0.1	7
mAb AS32 anti-phytochrome (IgG1)	CaMV 35S-ss-V_L-L-V_H L: EGKSSGSGSESKP	Tobacco PR protein (PR1a)	*N. tabacum*	apoplast and culture medium	0.25 - 0.5 µg/ml culture medium	16
mAb F8 anti-AMCV coat protein (IgG2b)	CaMV 35S-V_H-L-V_L L: GGGGSGGGGSGGGGS	none	*Nicotiana benthamiana*	N.D.	max 0.1	12
mAb NQ10-12.5 anti-2-phenyl-oxazol-5-one (phOx) (IgG)	LeB4-ss-V_H-L-V_L L: GGGGSGGGGSGGGGS	Legumin B4 (LeB4)	*N. tabacum*	secretory pathway	max 0.67	17
	LeB4-V_H-L-V_L L: GGGGSGGGGSGGGGS	none		N.D.	not detectable	
mAb 15-I-C5 anti-abscisic acid	CaMV 35S-ss-V_H-L-V_L L: GGGGSGGGGSGGGGS	Legumin B4 (LeB4) + KDEL	*N. tabacum*	endoplasmic reticulum and nuclear envelope	0.05 - 4.8	18

continued...

Table 9.2. Expression of antibody fragments in plants (continued)

Antibody	Construction (scFv / dAb)	Signal Sequence	Species	Localization	Expression Yield % of total soluble protein	Ref.
mAb 21C5 anti-cutinase of *Botrytis cinerea* (IgG1)	CaMV 35S-V_L-L-V_H-C_{H1}* *(only initial five triplets)	a) none	*N. tabacum*	N.D.	not detectable	93
	CaMV 35S-ss-V_L-L-V_H	b) a + KDEL		intracellular	max 0.2	
		c) mAb CEA66E3 κ chain		protoplast culture medium	max 0.01	
	L: REGKSSGSGSESKLEC	d) c + KDEL		intracellular	max 1	
mAb 6D4 anti-salivary secretion of root-knot nematode *Meloidogyne incognita* (IgM)	pT_R2'-V_L-L-V_H	a) none	*N. tabacum* protoplasts (transient expression)	intracellular	N.D.	94
	pT_R2'-ss-V_L-L-V_H L: EGKSSGSGSESKST (slightly modified)	b) a + KDEL		intracellular	N.D.	
		c) mAb CEA66E3 κ chain		intracellular	N.D.	
		d) c + KDEL		intracellular	N.D.	

mAb MAK 33 anti-human creatine kinase (IgG1)	CaMV 35S-V_L-L-V_H	none	*N. tabacum*	N.D.	max 0.01	95
	CaMV 35S-ss-V_L-L-V_H L: GGGGSGGGGSGGGGS	2S2 (*Arabidopsis thaliana* storage protein)		N.D.	max 0.01	
mAbs anti-BNYVV coat protein and anti-nonstructural protein (P25)	a)CaMV 35S-V_H-L-V_L	none	*N. benthamiana*	N.D.	not detectable	96
	b)CaMV 35S-ss-V_H-L-V_L	*Erwinia carotovora* pectate lyase (PelB)		intracellular (not secreted)	< 0.001	
	c)CaMV 35S-ss-V_H-L-V_L	*Phaseolus vulgaris* storage protein (PHA)		intracellular (not secreted)	0.002 - 0.01	
mAb NC1/34 HLanti-tachykynin substance P	CaMV 35S-ss-V_H	*E. carotovora* pectate lyase (PelB)	*N. benthamiana*	extracellular fluids	1	6

N.D.: not determined
L: linker peptide

of yeast), represents an example where plants differ from animal and yeast cells. Protein transport to the vacuole is dependent on specific signals to prevent secretion. The vacuolar targeting signals of several soluble plant proteins have been identified and three different types of sorting signals have been described: N-terminal and C-terminal propeptides (which are cleaved upon deposition of the protein in the vacuole) or regions of the mature protein.[65] Plant cells may have different types of vacuoles. Some vacuoles may derive directly from the ER and some storage proteins may be sorted directly from the ER to the vacuole. Moreover, sorting of proteins from Golgi to the vacuole implies another pathway for vacuolar compartmentation.[27]

Chloroplast ans Mitochondrial Transport

Other nonsecretory subcellular compartments are the chloroplasts, mitochondria, nucleus and peroxisomes. Due to the low coding capacity of the chloroplast and mitochondria genomes, most chloroplast and mitochondrial proteins are encoded in the nucleus and synthesized in the cytosol as larger precursors carrying an amino-terminal targeting signal (called 'transit peptide' for chloroplasts and 'presequence' for mitochondria), responsible for the correct import into the organelle. Often secondary targeting sequence information on the polypeptide ensures further movement to a different compartment within the organelle (e.g., thylakoids).

Although the mitochondrial targeting mechanism seems to be conserved between yeast and plants,[66] its characterization in plants has not been accomplished yet. On the contrary, chloroplast targeting has been better defined,[67] whereas the mitochondrial presequences seem to be conserved in plants, fungi and mammals,[68] the plastid-directing transit sequences are more heterogeneous in length and secondary structure. The common features of the chloroplast transit domains are an uncharged N-terminal region, a stromal processing site and a high content of serine and threonine residues.[69] This seems to indicate that protein import into the two organelles follows different routes.

It has been shown that a plant specific cytoplasmic protein kinase phosphorylates chloroplast-destined precursor proteins in an ATP-dependent manner, but not their mature forms nor mitochondrial and peroxisomal precursors.[70] On the other hand, mistargeting between chloroplasts and mitochondria have been observed in vivo and in vitro.[71,72]

Recently, combinations of mitochondrial and chloroplast targeting sequences were fused to the reporter genes β-glucuronidase (GUS) and chloramphenicol acetyl transferase (CAT).[73] Analysis of transgenic plants demonstrated that the presence of the chloroplast transit peptide alone was not sufficient to target the reporter proteins to the chloroplast. However, when the mitochondrial presequence was inserted downstream of the chloroplast sequence, import of the reporter proteins was observed. The amino-terminal position appeared to be fundamental in determining protein import specificity. In fact, the mitochondrial presequence alone was able to direct transport of CAT and to a lesser extent, GUS to mitochondria; GUS targeting to mitochondria was improved when the chloroplast targeting sequence was positioned after the mitochondrial presequence.

Nuclear Transport

Little is known about nuclear targeting in plants (for a review see ref. 74). Proteins that are destined to the nucleus are synthesized with short basic amino acid sequences, nuclear localization signals (NLSs) that drive them into the nucleus. Peculiar to plants, the moving in and out of some nuclear proteins is regulated by light.[75] Most studies have focused on NLSs from plant viruses[76] and bacteria.[77] Moreover, the SV40 large T-antigen NLS has been used to drive the RecA product into the plant nucleus demonstrating its functionality in plants.[78] Molecular components of the NLS binding site have been identified in plants[79] and plant nuclear import might have unique features with respect to the vertebrate and yeast systems.[80]

Targeting to Peroxisome

Peroxisomes are ubiquitous intracellular organelles, (involved in oxidizing processes) which unlike chloroplast and mitochondria, posses a unique membrane. Proteins that are destined to peroxisomes are first synthesized in the cytoplasm and then imported into the organelle. Two peroxisomal targeting signals have been identified[81] (PTS1, noncleavable C-terminal SKL-like tripeptide, and PTS2, cleavable N-terminal). Chimeric proteins fused to the PTS-like sequences have been correctly directed into this compartment.[82]

In conclusion, the mechanisms controlling plant protein folding, assembly and transport seem to not be too different from those of mammalian cells, but specific and distinct signals are used. Hence, theoretically, expression of antibodies in different compartments of plant cells could be extended similarly to what has been performed in animal cells.

Antibody Assembly and Secretion

Several groups have reported high expression of functional full-length antibodies in the secretory pathway of transgenic plants (see Table 9.1). Most of these "plantibodies" recognize and bind the antigen with similar affinity and specificity as the parental antibody, demonstrating a correct three-dimensional folding. Expression in plants of each immunoglobulin chain alone resulted in significantly lower (30-fold) accumulation levels than when both chains were expressed simultaneously.[2] In the latter case, all immunoglobulin chains expressed were assembled into heavy-light chain complexes.[2] Furthermore, Voss et al[83] detected equimolar amounts of both immunoglobulin chains in transgenic plants simultaneously expressing both chains, although light chain transcripts were more abundant than those for the heavy chain. These results indicate an instability of the unassembled chains. Expression of whole antibodies in the ER/apoplast of plants contrasts the unsuccessful attempts to express complete antibodies in the bacterial periplasm, in which only the expression of smaller antibody fragments has been successful. This difference may be due to the absence of glycosylation in bacteria since addition of carbohydrate moieties may contribute greatly to stability.[84] The efficiency of immunoglobulin chain assembly in the transgenic plants obtained so far is subject to high variations which may depend on the transformation strategy and the promoters used. Sexual crossing of individually-transformed plants expressing light or heavy chains results in higher assembly efficiency.[85] Simultaneous transformation (cotransformation) with two vectors bearing heavy and light constructs induced a lower percentage of assembly and a high variability in the assembly efficiency among the various transformants.[15] This may be due to an imbalance in the expression of the two chains. Coordinated promoters induced high levels of assembled antibody[14,85] while no antibody could be expressed after sexual crossing of two plants expressing immunoglobulin transgenes, one under the control of the constitutive 35S promoter and the other under the heat shock inducible promoter HSP70.[85] This may be due to different cellular expression of the individual chains. Variations in assembly efficiency were found among different species. Antibody assembly efficiency in *Arabidopsis* is 3-fold higher than in *Nicotiana*.[15] The reasons for this are unclear but it is possible that the plant-species differ in the efficiency of immunoglobulin assembly/folding.

To drive immunoglobulins in the secretory compartment, signal sequences from mouse, plant, yeast and bacteria have been used (Table 9.1) and demonstrates the origin of the signal sequence does not appear to be critical. However, secretion has only been formally demonstrated in very few cases. Hein et al[13] showed secretion across the plant cell membrane and cell wall using pulse chase labeling experiments with protoplasts and cell suspensions from transgenic plants. In this study, correct processing of the signal peptide was confirmed by light chain N-terminal sequencing. DeWilde et al[86] were recently able to immunolocalize assembled full length IgG molecules in the intercellular spaces of transgenic *Arabidopsis* plants. This clearly suggests that the immunoglobulin is not physically restricted from passage through the plant cell wall, notwithstanding an estimated exclusion limit for globular proteins of 17 kDa[87] or 40-60 kDa.[88] Should this property be ascribed to all full length antibodies, the possibility to restrain extracellular antigens (intercellular immunization) could be exploited. Conversely, secretory IgA molecules seem not to permeate the cell wall.[20] Surprising targeting patterns have been found in tobacco plants expressing a full length antibody.[4] In this case the native signal sequence was substituted by a signal sequence deriving from barley α-amylase. Assembled antibodies were detected within the ER, but not further downstream in the secretory system of transgenic tobacco plants, furthermore, immunoglobulin chains were mistargeted to the chloroplasts, although the same α-amylase signal peptide was able to drive a chimeric T4 lysozyme to the intercellular spaces of transgenic tobacco tissues.[89] In addition, Voss et al[83] mentioned the unsuccessful use of the same α-amylase signal peptide which resulted in only one transgenic plant expressing very low levels of antibodies. There is the possibility that the monocot-derived α-amylase signal sequence is not suitable for full-length antibody expression in plants. Problems were also encountered with the signal sequence of the 2S2 storage protein of *Arabidopsis*[90] which guided the expression of Fab molecules to the nucleolus of tobacco meristematic shoot tissue.[15]

Antibody Derived Fragments

Expression in plants of either a leader-less heavy or light chain constructs, resulted in very poor yields of each chain if compared to constructs equipped with signal sequence (40-fold and 20-fold less, respectively).[2] However, Northern analysis showed the presence of nearly equivalent amounts of two transcripts, demonstrating instability of the immunoglobulin chains in the cytoplasm. In filial recombinants of transgenic plants coexpressing leader-less immunoglobulin chains, no functional assembled antibody was detected unlike what was found in animal cells[105-107] (see chapter 7). This may be ascribed to the fact that the cytoplasm is a reducing environment[91] and consequently, the formation of disulfide bonds which is crucial for antibody domains stability and folding is not favored. Furthermore, the higher amount of proteases in the cytosol in comparison to the ER/apoplast may have contributed to the lack of cytosolic expression of whole antibody in plant cells. Contrary to the higher plants, the alga *Acetabularia mediterranea*[92] allowed the expression of assembled complete antibody inside the cytoplasm. This report lacks functional data. If indeed this assembled antibody is also functional, then it is possible that the cytoplasm of the alga is less reducing than that of the higher plants. Alternatively, differences in the proteases present in different plant species and in animal cells may account for this variability.

In principle, ScFv fragments minimize problems in assembly since the presence of both variable domains in a single chain ensures their balanced expression. Expression of secretory ScFv in the plant apoplast[16,93-95] resulted in high accumulation levels (0.1%-1%), matching those of a complete antibody. Interestingly, scFv also can be stably and functionally expressed in the cytoplasm of the plants[7,12,94] by using leader-less constructs with expression levels varying from 0.06% to 0.1%. However, this does not seem to hold true for all scFv antibodies. Several scFvs with different specificities failed to demonstrate functional expression in the cytoplasm of the plants[17,93,95,96] although the corresponding mRNAs have always been detected. Furthermore, Owen et al[16] reported that a massive screening of putative transgenic plants was necessary to find a plant showing significant expression levels (0.06% total soluble proteins) of cytoplasmic scFv, while only few of plants were screened to isolate good producers (0.6% total soluble proteins) of the secretory version of the same scFv.[16] Also in this case, equal mRNA levels have been detected. Although differences in the translatability of the two transcripts cannot be excluded, these results demonstrate a reduced scFv stability in the cytoplasm. On the contrary, in the case of cytoplasmic anti-viral scFv(F8),[12] several independent transformants have been obtained with high expression levels. Furthermore, in vivo and in vitro studies indicated high stability of this particular ScFv (our unpublished results). The factors determining the differences in the stability between the various scFvs in the plant cytoplasm are still not clear but most are probably linked to the primary sequence (see chapter 4). Only antibody molecules having sufficient free energy of folding may tolerate the lack of the stabilizing effect of the disulfide bonds and of the folding catalysts. Residues in the framework of the antibody domains that may increase stability of the antibody molecule expressed in plant are not known. It is interesting to verify whether residues found to improve stability[97] and folding[98] of an antibody fragment in bacteria may also increase stability of various antibody fragments in the plant cytoplasm or whether the inefficient cytoplasmic disulfide formation is the limiting step in the folding pathway.

Studies on the redox state of the scFv(F8) in the plant cytosol indicate that this antibody fragment accumulates in the cytoplasm in a reduced form (manuscript in preparation), as in the cytosol of mammalian cells,[106] suggesting that the formation of the disulfide bond is not the only determinant of correct antibody folding/assembly. This is in agreement with the fact that overexpression of the *E. coli* disulfide isomerase DsbA in the bacterial periplasm did not increase the antibody yield.[99] Most probably aggregation processes, occurring before the action of the disulfide isomerase through folding intermediates, determine the folding efficiency (see chapter 4). It is possible that the primary structure determines the aggregation processes. Furthermore, we cannot exclude the possibility that the cytosolic chaperonins[52] recognize and bind differentially the various antibodies. In addition, the presence within the plant cytosol of proteases that may specifically or preferentially recognize some antibodies, but not others, could be also responsible for the unsuccessful attempts to express some scFv antibodies in the plant cytoplasm. In this regard, the high proteolytic instability of some linker peptides used for the construction of scFv antibodies needs to be taken in consideration.[100-103]

Although the leader-less scFv(F8) accumulates in the reduced form, the possibility of mistranslocation to ER or different subcellular compartments cannot be excluded, considering what was reported for animal cells.[104] In addition, the possibility of stabilization of a cytosolic scFv through interaction with the cytosolic face of membranes should also be considered. A possibility has been also mentioned that antigen-antibody interaction may increase the stability of antibody molecules.[106-108] However, this does not suit for the anti-viral scFv(F8), as no significant difference in the ScFv antibody accumulation has been detected in transgenic plants uponvital infection (unpublished results).

Several signals have been used to drive the expression of ScFv molecule in the secretory pathway (Table 9.2). Retention of a secretory scFv in the ER through a KDEL sequence resulted in an increased amount of protein.[93,94] In another case, expression of secretory ScFv in transgenic plants was only possible when the KDEL sequence has been included in the expression cassette.[18] These differences in the accumulation levels of the scFv protein in the presence or absence of the KDEL motif are not accompanied by differences in the accumulation of the corresponding mRNAs, indicating different protein stability. It is possible that localization in the ER protects the molecule from proteolytic activity further down the secretory pathway. Interestingly enough, the KDEL motif confers stability also to the leaderless version of the ScFv construct.[93,94] It is possible that the KDEL sequence may somehow protect the cytosolic ScFv from proteolytic degradation. Alternatively, it is possible that the leader-less ScFv normally located in the cytosol is targeted to the ER by a cryptic N-terminal leader sequence.

Immunotherapy Against Plant Diseases

Like animals, plants are continuously exposed to pathogen attack and the range of phytopathogenic organisms is diverse, including viruses, mycoplasmas, bacteria, fungi, protozoa, nematodes and other parasites. Lacking a true immune system, plants have evolved defense mechanisms which are completely different from vertebrates. Each cell is capable of defending itself and resistance to a pathogen is often correlated with the so called 'hypersensitive response', an induced localized cell death at the site of infection that blocks further invasion. Plant breeding has utilized for many years wild relatives of crop plants as a source of resistance genes, but the recurrent scarcity of natural occurring resistance-genes generated a great interest towards the use of molecular methods to protect plants from diseases. Recently, some resistance genes involved in the mechanisms of recognition and activation of the signal transduction pathway, which leads to the 'hypersensitive response' have been cloned.[109] The unraveling of their basic mechanisms of action will possibly allow in the future, the development of novel methods of disease control. In a complementary and alternative approach, antibodies may provide a rich source for potential resistance genes.

Genetic transformation to obtain plants resistant to pathogen has been an early target of agricultural biotechnology and several strategies have been developed, mainly against viral diseases. Protection to viruses was mainly achieved by the expression of viral sequences (i.e. coat protein, movement protein, replicase) as well as antisense, satellite and defective interfering RNA molecules, generally referred as to as 'pathogen-derived resistance.'[110] However, debate is now open on whether viral transgenes may imply certain environmental risks[111] being the starting point for the evolution of novel virus variants.[112]

Although plants do not use antibodies as part of their defense arsenal, the unconventional idea to exploit the vertebrate immune system to endow plants with new forms of disease defense,[6,12] is now providing the basis for a revolutionary 'immunotherapy'. Immunotherapy seems to be the ideal approach for viral pathogens because of the relative simplicity of the viral genome which allows molecular dissection of their genetic components. Viral proteins responsible for a defined pathogenic function can be used as targets for developing antibodies able to perturb fundamental steps in the viral life cycle. Moreover, the plantibody strategy based on the high specificity of antigen-antibody interaction, represents a measure of virus biocontrol devoid of recombination/mutation risks and should not interfere with host molecules involved in physiological processes.

The feasibility of plant immunoprotection against virus has been first reported by Tavladoraki et al.[12] A constitutively expressed cytoplasmic ScFv raised against the coat protein of AMCV (artichoke mottle crinkle virus) was able to specifically shield transgenic plants from viral attack, reducing the infection incidence and causing a delay in symptoms development upon challenge with abnormally high titers of viral inocula. Fundamental for the protection was the expression of this molecule in the cytoplasm where it can interfere with the early steps of viral infection (see Fig. 9.1). In a similar approach, reduced infectivity was achieved in tobacco plants expressing secretory full length monoclonal antibody raised against a conserved epitope of TMV (tobacco mosaic virus) coat protein.[83] As infection of TMV occurs only after wounding, the basic idea was to export antibodies to the intercellular space where large amount of immunoglobulins supposedly encounter the virus before entering the cell. It is worthwhile noting that all available evidence suggest that in plants, no virus specific receptors are present on the plasma membrane and no clear mechanism is known that accounts for virus uncoating. Whatever the mechanisms, if an antibody is directed towards a structural protein like the coat protein, domains involved in intersubunit interactions stabilizing the capsid may represent vulnerable targets.[12] In this way, events like disassembly and assembly are likely to be impaired. Such an antibody may also block the movement of some viruses, as coat protein is believed to play a role either in long distance spread of some viruses through the phloem or in cell-to-cell movement through plasmodesmata.[113] However, it is important to consider that the viral capsid is a multimeric antigen, and neutralization may require molar excess of antibody.[114]

Another possible structural target is the movement protein, responsible for cell-to-cell movement of the virus through plasmodesmata.[115] Directing an antibody to the movement protein may prevent the typical changes of plasmodesmata determined by this protein[116] and restrict the virus at the original site of infection.

Targeting the catalytic site of enzymatic nonstructural viral proteins (replicase, helicase, protease) may be even more efficient than targeting coat or movement proteins. Furthermore, replicases (RNA-dependent RNA polymerases) from

Fig. 9.1. Effect of artichoke mottle crinkle virus (AMCV) on transgenic and control *Nicotiana benthamiana* plants 15 days after virus challenge (10^8-10^9 particles per plant). Left: Transgenic plant expressing cytoplasmic ScFv raised against the coat protein of AMCV. Right: Untransformed control.

a wide variety of positive-strand RNA viruses share several domains of sequence similarity.[117,118] These highly conserved motifs are probably less susceptible to mutation and an antibody recognizing these regions might provide plants with durable protection against a wide range of evolutionarily distant viruses. It would also be interesting to investigate if an antibody able to inhibit virus-encoded proteases, which are responsible for processing viral polyproteins, will be able to block virus spread throughout the plant.

Preventing nematode infection is one area where the immunotherapy may prove effective. Plant parasitic nematodes colonize plant cells through a hollow stylet (necessary to pierce the cell wall) inducing dramatic changes in the plant cell metabolism triggered by not yet identified signals contained in their salivary secretions.[119] Blocking the development feeding site through the expression of a full length antibody specific to stylet secretions, has been attempted recently.[120] In this early attempt, the authors reported no interference of the antibody in the invasive process. This has been ascribed to the fact that secretions are released in the cytoplasm while the antibody accumulated in the apoplast. The same antibody engineered in the scFv format and transiently expressed, successfully accumulated in the cytoplasm of tobacco protoplasts.[94] It would be interesting to verify whether this molecule, once expressed in the cytoplasm of transgenic plants, could hinder root-knot nematode infection. It appears that fundamental for effective immunotherapy is the knowledge of the relevant compartment where the antibody should restrain the antigen, possibly when concentration of the antigen is still low. This general concept applies to any plant pathogen.

A great number of fungi of the major phylogenetic groups causes severe diseases in plants. During the early events of infection, fungi secretes a mixture of hydrolytic and proteolytic enzymes. In the case of some fungi, interference during the early event of plant colonization process can be obtained through secretion of antibodies against specific pathogenic factors into the cell wall/intercellular spaces. Favorite targets are the cell wall degrading enzymes such as polygalacturonases and/or pectate lyases; but the picture is even more complicated as fungi have evolved complex strategies to suppress and avoid host defense which imply the switching on/off of different pathogenicity factors.[121]

Albeit genetic immunization of plants could hypothetically be applied to any disease, it appears clear that successful applications demand a deep comprehension of specific pathogenic functions to be hampered, together with the knowledge of stability, kinetics, affinity of each individual antibody.

Immunotherapeutic Plantibodies: Biomedical Applications

The capability of plant cells to produce full-length antibodies and the flexibility to with which the immunoglobulin domain structure can be adapted to different experimental designs, positively influenced the exploitation of transgenic plants as sources of antibodies to be used in immunotherapy. Although the prospect of developing therapeutic monoclonal antibodies is rather complex, irrespective of the origin of the expression system used, plants are particularly advantageous when large quantities of antibody are needed, such as in the case when antibodies are the source of therapeutic antibodies, for passive immunization. In fact, topical immunotherapy seems a promising area where the production of plantibodies has been explored successfully.

In the field of dental caries protection in humans, early studies demonstrated that in nonhuman primates, the colonization of teeth by the caries principal causative agent, *Streptococcus mutans,* was prevented by a monoclonal antibody (Guy's13) with specificity to an epitope of the streptococcal adhesin SA I/II.[19] In parallel, the topical application of this mAb was able to grant a long-lasting protection to humans. With the idea that such a mAb could be more acceptable as a therapeutic if produced by plants, the genes encoding the heavy and light chain were engineered for stable expression in plants. As an alternative, the α domains of an IgA antibody were introduced in the constant region giving rise to a chimeric IgG-IgA molecule that has been assembled in plants without detrimental effect to stability and binding.[19] These findings were fundamental to the successive production of the dimeric secretory IgA in plants, the prevailing immunoglobulin in mammal mucosal secretions, where it exerts a protective function against infectious agents. To be secreted in the mucosal environment, the IgA molecule dimerized by the J chain is secreted by plasma cells and is further subjected to a receptor mediated endocytosis by epithelial cells that produce the secretory component (SC). This polypeptide binds the dimeric IgA initiating the complex process of secretion from epithelium, conferring a high degree of resistance to proteolysis in the severe conditions of the gastrointestinal tract. The production of monoclonal secretory IgA would be of fundamental importance for immunotherapy, but achieving sufficient amounts of this molecule is difficult due to the fact that two different cell types need to participate in the synthesis. On the contrary, transgenic plants proved to be able to assemble large quantities of secretory IgA.[20] After the evidence that the chimeric IgG-IgA molecule is functionally expressed in tobacco plants, the secretory dimerized version was completed employing four transgenic tobacco plants to express a murine κ chain, the hybrid IgA-IgG heavy chain, the murine J chain and a rabbit secretory component (SC). Following successive crossing between these plants, a single transgenic plant resulted that was expressing at the same time, the four polypeptide chains. The assembly of all components was successfully achieved and the yield of secretory immunoglobulin A was extremely satisfactory (see Table 9.1). These results demonstrate that heterologous complex molecules such as secretory IgA can be expressed by plants and passive immunization of all mucosal surfaces can be foreseen, especially through edible plant products (tubers, fruit, seeds), that may eventually eliminate the need to purify the plantibody for immunotherapy.

In general, one of the limitations of immunotherapy of some human and animal diseases is that large amounts of specific antibody are required for the treatments. It has been estimated that plantibodies expressed at the level of 1% of total soluble proteins could be produced at a cost of US $100 per kilogram.[122] Hence expectations of utilizing plants as biofactories of valuable antibodies hold encouraging results from both the economical and the technical viewpoint. Whether or not the agricultural scale production will actually take place is strictly dependent on specific safety issues that will be addressed for these plant derived antibodies. This is an entirely new field as no clinical trial or animal applications have been performed yet, that make use of plantibodies. It is conceivable that the plant derived antibodies for human and animal applications will be requested to fulfill the safety and performance standards of antibodies derived from other nonmammalian systems like yeasts or bacteria.

One issue of concern that the production of complete antibody in plants should take into account is the glycosylation pattern which may be distinct from that of animal cells. In fact, the complex oligosaccharides generated during transport through the Golgi could consist of xylose, fucose and/or N-acetylglucosamine. This does not influence antigen binding nor specificity, hence the addition of some carbohydrate moieties unique to plants might affect only the immunogenicity of the plant produced antibody, especially if administered systemically. This prospect requires a careful evaluation, depending upon the experimental design for the recombinant antibody, although this issue is clearly important for all production systems and not only for plants. Conversely, in the case of food derived from transgenic plants, the risks for animal and human health are likely to be negligible, considering the frequent exposure to immunoglobulins and/or plant glycoproteins in food and the absence of receptors for immunoglobulin in the adult mammalian gut. Any risk of this sort becomes even more unlikely when using engineered antibodies that lack the Fc region.

Perspectives and Conclusions

So far antibody production in plants extends from the expression of minimal binding units to large dimeric molecules offering exciting perspectives into two major fields: plant immunomodulation and large scale production for human therapy. For plant immunomodulation, ideal molecules are those amenable to easy expression and assembly. The prevalent opinion is that expression of full antibody molecules in plants is troublesome because of the requirement to transfer two gene constructs and complexity in assembly and folding. On the contrary, most of the engineered scFvs do not give particular problems in assembly/folding and can even be stably accumulated in the plant cytoplasm.[7,12,94] In this way, any intracellular target may be reached. Furthermore, the small size of scFv molecules may allow them easy access to sites on the target molecules that are inaccessible to full-sized immunoglobulin. When the target is a multimeric protein, the relatively small scFv molecule may bind to a larger number of sites than the parental antibody would be able to. Conversely, the stability and affinity of the engineered scFv compared to that of the parental antibody may vary depending mostly on the linker peptide connecting the two variable domains[101] their orientation in the scFv construct[123] and the primary sequence of the antibody.[98] Hence the major limiting factors are those influencing the physicochemical parameters associated with engineered immunoglobulin fragments. Remarkably, the scFv antibody may provide the scaffold for the construction of either bivalent or bispecific antibody.[124,125] The value of these molecules lies in their avidity, which may be greater than that of the monovalent antibody, and their potential ability to retain two targets. A bifunctional scFv in which the antibody is fused to a toxin[126] may also improve the efficacy of the scFv strategy in inhibiting virus replication and spreading, but also an elective system for easy production of immunoconjugate of medical interest due to the low productive efficiency of some toxins in heterologous nonplant hosts. Bivalent and bispecific antibodies may provide another class of highly active molecules, yet unexplored for immunoprotection of plants.

Finally, the comparison with other heterologous systems highlights the long procedure to obtain transgenic plants for the bulk production of antibodies intended for human or veterinary therapeutic uses. While in the case of tobacco two to three months are necessary to complete the entire transgenic plant cycle, procedures could be even longer with edible plants. Consequently, although plants potentially are able to express virtually any kind of antibody, there must be a definite requirement for selecting plants as hosts for bulk production. In fact, traditional biofermentation techniques have highly evolved and remarkable yields have been described in bacteria and yeasts. In this view, the use of plants as bioreactors for bulk production of antibodies for diagnostic uses should be carefully evaluated. It is prevalent opinion that plants retain a unique role for bulk production of antibodies of high value for passive immunotherapy; therefore, if the questions associated to glycosylation can be experimentally evaluated and addressed, many exciting biomedical applications can be envisaged in this area.

It is important to notice that the use of antibodies in plants for both plant immunomodulation and animal immunotherapy requires new insights as to how the plant cell machinery can process this evolutionarily distant protein. The progress made in a few years time forecasts the plantibody technology may soon be in the position of answering the unresolved questions and of extending the concept of immunotherapy to a broader sense.

References

1. Moffat AS. Exploring transgenic plants as a new vaccine source. Science 1995; 268:658-660.
2. Hiatt A, Cafferkey R, Bowdish K. Production of antibodies in transgenic plants. Nature 1989; 342:76-78.
3. Sheng J, Citovsky V. *Agrobacterium*-plant cell DNA transport: have virulence proteins, will travel. Plant Cell 1996; 8:1699-1710.
4. Düring K, Hippe S, Kreuzaler F et al. Synthesis of a functional monoclonal antibody in transgenic *Nicotiana tabacum*. Plant Mol Biol 1990; 15:281-293.
5. Winter G, Milstein C. Man-made antibodies. Nature 1991; 349:293-299.
6. Benvenuto E, Ordàs R, Tavazza R et al. 'Phytoantibodies': a general vector for the expression of immunoglobulin domains in transgenic plants. Plant Mol Biol 1991; 17:865-874.

7. Owen M, Gandecha A, Cockburn B et al. Synthesis of a functional anti-phyto-chrome single-chain Fv protein in transgenic tobacco. Bio/technology 1992; 10:790-794.

8. Ward ES, Gussow D, Griffiths AD et al. Binding activities of a repertoire of single immunoglobulin variable domains secreted from *E. coli*. Nature 1989; 341:544-546.

9. Hamers-Castermann C, Atarhouch T, Muyldermans S et al. Naturally occurring antibodies devoid of light chains. Nature 1993; 363:446-448.

10. Desmyter A, Transue TR, Ghahroudi MA et al. Crystal structure of a camel single-domain VH antibody fragment in complex with lysozyme. Nature Struct Biology 1996; 3:803-811.

11. Spinelli S, Frenken L, Bourgeois D et al. The crystal structure of a llama heavy chain variable domain. Nature Struct Biology 1996; 3:752-757.

12. Tavladoraki P, Benvenuto E, Trinca et al. Transgenic plants expressing a functional single-chain Fv antibody are specifically protected from virus attack. Nature 1993; 366:469-472.

13. Hein MB, Tang Y, McLeod DA et al. Evaluation of immunoglobulins from plant cells. Biotechnol Prog 1991; 7:455-461.

14. van Engelen FA, Schouten A, Molthoff JW et al. Coordinate expression of antibody subunit genes yields high levels of functional antibodies in roots of transgenic tobacco. Plant Mol Biol 1994; 26:1701-1710.

15. De Neve M, De Loose M, Jacobs A et al. Assembly of an antibody and its derived antibody fragment in *Nicotiana* and *Arabidopsis*. Transgenic Research 1993; 2:227-237.

16. Firek S, Draper J, Owen MRL et al. Secretion of a functional single-chain Fv protein in transgenic tobacco plants and cell suspension cultures. Plant Mol Biol 1993; 23:861.

17. Fiedler U, Conrad U. High-level production and long-term storage of engineered antibodies in transgenic tobacco seeds. Bio/technology 1995; 13:1090-1093.

18. Artsaenko O, Peisker M, zur Nieden U et al. Expression of a single-chain antibody against abscissic acid creates a wilty phenotype in transgenic tobacco. The Plant J 1995; 8:745-750.

19. Ma JK-C, Lehner T, Stabila P et al. Assembly of monoclonal antibodies with IgG1 and IgA heavy chain domains in transgenic tobacco plants. Eur J Immunol 1994; 24:131-138.

20. Ma JK-C, Hiatt A, Hein M et al. Generation and assembly of secretory antibodies in plants. Science 1995; 268:716-719.

21. Bar-Peled M, Bassham DC, Raikel NV. Transport of proteins in eukaryotic cells: more questions ahead. Plant Mol Biol 1996; 32:223-249.

22. Chrispeels MJ. Sorting of proteins in the secretory system. Annu Rev Plant Mol Biol 1991; 42:35-49.

23. Bednarek SY, Raikel NV. Intracellular trafficking of secretory proteins. Plant Mol Biol 1992; 20:133-150.

24. Vitale A, Ceriotti A, Denecke J. The role of the endoplasmic reticulum in protein synthesis, modification and intracellular transport. J Exp Bot 1993; 44:1417-1444.

25. Lucas WJ, Wolf S. Plasmodesmata: the intercellular organelles of green plants. Trends Cell Biol 1993; 3:308-15.

26. Vitale A, Chrispeels MJ. Sorting of proteins to the vacuoles of plant cells. BioEssays 1992; 14:151-60.

27. Okita TW, Rogers JC. Compartmentation of proteins in the endomembrane system of plant cells. Ann Rev Plant Physiol Plant Mol Biol 1996; 47:327-50.

28. Walter P, Lingappa V. Mechanisms of protein translocation across the endoplasmic reticulum. Annu Rev Cell Biol 1986; 2:499-516.

29. Prehn S, Wiedmann M, Rapoport TA et al. EMBO J. 1987; 6:2093-2097.

30. Miernick JA, Shatters RG. The use of maize endosperm microsomes for analysis of translocation and processing of secretory precursors. Plant Physiol 1992; 99 Suppl: 44.

31. Sijmons PC, Dekker BMM, Schrammeijer B et al. Production of correctly processed human serum albumin in transgenic plants. Bio/technology 1990; 8:217-221.

32. Denecke J, Botterman J, Deblaere R. Protein secretion in plant cells can occur via a default pathway. Plant Cell 1990; 2:51-59.

33. Lund P, Lee RY, Dunsmuir P. Bacterial chitinase is modified and secreted in transgenic tobacco. Plant Physiol 1989; 91:130-135.

34. Hunt DC, Chrispeels MJ. The signal peptide of a vacuolar protein is necessary and sufficient for the efficient secretion of a cytosolic protein. Plant Physiol 1991; 96:18-25.

35. Pfeffer SR, Rothman JE. Biosynthetic protein transport and sorting by the endoplasmic reticulum and golgi. Annu Rev Biochem 1987; 56:829-852.

36. Denecke J, De Rycke R, Botterman J. Plant and mammalian sorting signals for protein retention in the endoplasmic reticulum contain a conserved epitope. EMBO J 1992; 11:2345-2355.

37. Pelham HRB. Evidence that luminal ER proteins are sorted from secreted proteins in a post-ER compartment. EMBO J 1988; 7:913-918.

38. Sönnischen B, Füllekrug J, van Nguyen P, et al. Retention and retrieval: both mechanism cooperate to mantain calreticulin in the endoplasmic reticulum. J Cell Sci 1994; 107:2705-2717.

39. Bar-Peled M, Conceiçao AS, Frigerio et al. Expression and regulation of a ERD2, a gene encoding the KDEL receptor homolog in plants and other genes encoding proteins involved in ER-Golgi vesicular trafficking. Plant Cell 1995; 7:667-676.

40. Munro S, Pelham HRB. A C-terminal signal prevents secretion of luminal ER proteins. Cell 1987; 48:899-907.

41. Herman EM, Tague BW, Hoffman LM et al. Retention of phytohemagglutinin with carboxyterminal tetrapeptide KDEL in the nuclear envelope and the endoplasmic reticulum. Planta 1990; 182:305-312.

42. Okamoto T, Nakayama H, Seta K et al. Posttranslational processing of a carboxy-terminal propeptide containing a KDEL sequence of plant vacuolar cysteine endopeptidase (SH-EP). FEBS Lett 1994; 351:31-34.

43. d'Enfert C, Gensse M, Gaillardin C. Fission yeast and a plant have functional homologues of the Sar1 and Sec12 proteins involved in the ER to Golgi traffic in budding yeast. EMBO J 1992; 11:4205-4211.

44. Bassham DC, Gal S, Conceicao AS et al. An *Arabidopsis* synthaxin homologue isolated by functional complementation of a yeast *pep12* mutant. Proc Natl Acad Sci 1995; 92:7262-7266.

45. Bassham DC, Raikel NV. Transport proteins in the plasma membrane and the secretory system. Trends Plant Sci 1996; 1:1520.

46. Gal S, Raikel NV. Protein sorting in the endomembrane system of plant cells. Curr Opin Cell Biol 1993; 5:636-640.

47. Hartl FU. Molecular chaperones in cellular protein folding. Nature 1996; 381:571-580.

48. Fontes EBP, Shank BB, Wrobel RL et al. Characterization of an immunoglobulin binding protein homolog in the maize *floury-2* endosperm mutant. Plant Cell 1991; 3:483-96.

49. Walther-Larsen H, Brandt J, Collinge DB et al. A pathogen-induced gene of barley encodes a HSP90 homologue showing striking similarities to vertebrate forms resident in the endoplasmic reticulum. Plant Mol Biol 1993; 21:1097-1108.

50. Huang L, Franklin AE, Hoffman NE. Primary structure and characterization of an *Arabidopsis thaliana* calnexin-like protein. J Biol Chem 1993; 268:6560-6566.
51. Chen F, Hayes PM, Mulrooney DM et al. Identification and characterization of cDNA clones encoding plant calreticulin in barley. Plant Cell 1994; 6:835-843.
52. Boston RS, Viitanen PV, Vierling E. Molecular chaperones and protein folding in plants. Plant Mol Biol 1996; 32:191-222.
53. Denecke J, Goldman MHS, Demolder J et al. The tobacco luminal binding protein is encoded by a multigene family. Plant Cell 1991; 3:1025-35.
54. Denecke J, Carlsson LE, Vidal S et al. The tobacco homolog of mammalian calreticulin is present in protein complexes in vivo. Plant Cell 1995; 7:391-406.
55. D'Amico L, Valsasina B, Daminati MG et al. Bean homologs of the mammalian glucose-regulated proteins: induction by tunicamiycin and interaction with newly synthesized seed storage proteins in the endoplasmic reticulum. Plant J 1992; 2:443-55.
56. Pedrazzini E, Giovinazzo G, Bollini R et al. Binding of BiP to an assembly-defective protein in plant cells. Plant J 1994; 5:103-10.
57. Hammond C, Helenius A. Quality control in the secretory system. Curr Opin Cell Biol 1995; 7:523-529.
58. Li X, Wu Y, Zhang D-Z et al. Rice prolamine protein body biogenesis: a Bip-mediated process. Science 1993; 262:1054-1056.
59. Shimoni Y, Segal G, Zhu X et al. Nucleotide sequence of a wheat cDNA encoding protein disulfide isomerase. Plant Physiol 1995; 107:281.
60. Li C-P, Larkins BA. Expression of protein disulfide isomerase is elevated in the endosperm of the maize *fluory*-2 mutant. Plant Mol Biol 1996; 30:873-882.
61. Gething M-J, Sambrook J. Protein folding in the cell. Nature 1992; 355:33-45.
62. Staehelin LA, Moore I. The plant Golgi apparatus: structure, functional organization and trafficking mechanisms. Annu Rev Plant Physiol Plant Mol Biol 1995; 46:261-288.
63. Fiedler K, Simons K. The role of N-glycans in the secretory pathway. Cell 1995; 81:309-312.
64. Sturm A, Kuick AV, Vliegenthart JFG et al. Structure, position and biosynthesis of the high mannose and the complex oligosaccharide side chains of the bean storage protein phaseolin. J Biol Chem 1987; 262:13392-13403.
65. Nakamura K, Matsuoka K. Protein targeting to the vacuole in plant cells. Plant Physiol 1993; 101:1-5.
66. Chaumont F, O'Riordan V, Boutry M. Protein transport into mitochondria is conserved between plant and yeast species. J Biol Chem 1990; 265:16856-16862.
67. de Boer AD, Weisbeek PJ. Chloroplast protein topogenesis, import, sorting and assembly. Biochim Biophys Acta 1991; 1071:221-253.
68. Hartl F-U, Neupert W. Protein sorting to mitochondria; evolutionary conservations of folding and assembly. Science 1990; 347:930-938.
69. von Heijne G, Nishikawa K. Chloroplast transit peptides. The perfect random coil? FEBS Lett 1991; 278:1-3.
70. Waegemann K, Soll J. Phosphorylation of the transit sequence of chloroplast precursor protein. J Biol Chem 1996; 271:6545-6554.
71. Brink S, Flugge UI, Chaumont F et al. Preproteins of chloroplast envelope inner membrane contain targeting information for receptor-dependent import into fungal mitochondria. J Biol Chem 1994; 2696:16478-16485.
72. Huang J, Hack E, Thornburg RW et al. A yeast mitochondrial leader peptide functions in vivo as a dual targeting signal for both chloroplast and mitochondria. Plant Cell 1990; 2:1249-1260.

73. Silva Filho M de C, Chaumont F, Leterme S et al. Mitochondrial and chloroplast targeting sequences in tandem modify protein import specificity in plant organelles. Plant Mol Biol 1996; 30:769-780.
74. Raikel NV. Nuclear targeting in plants. Plant Physiol 1992; 100:1627-1632.
75. Deng X-W. Fresh view of light signal transduction in plants. Cell 1994; 76:423-426.
76. Carrington JC, Freed DD & Leinicke AJ. Bipartite signal sequence mediates nuclear translocation of the plant potyviral NIa protein. Plant Cell 1991; 3:953-962.
77. Citovsky V & Zambryski P. Transport of nucleic acids through membrane channels: snaking through small holes. Annu Rev Microbiol 1993; 47:167-197.
78. Reiss B, Klemm M, Kosak H et al. RecA protein stimulates homologous recombination in plants. Proc Natl Acad Sci 1996; 93:3094-3098.
79. Hicks GR, Raikel NV. Nuclear localization signal binding proteins in higher plant nuclei. Proc Natl Acad Sci 1995; 92:734-738.
80. Hicks GR, Smith HMS, Lobreaux S et al. Nuclear import in permeabilized protoplasts from higher plants has unique features. Plant Cell 1996; 8:1337-1352.
81. Gould SJ, Keller GA, Hosken N et al. A conserved tripeptide sorts proteins to peroxisomes. J Cell Biol 1989; 108:1657-1664.
82. Olsen LJ, Ettinger WF, Damsz B et al. Targeting of glyoxysomal proteins to peroxisomes in leaves and roots of a higher plant. FEBS Lett 1993; 5:941-952.
83. Voss A, Niersbach M, Hain R et al. Reduced virus infectivity in *N. tabacum* secreting a TMV-specific full-size antibody. Mol Breeding 1995; 1:39-50.
84. Shin SU, Wright A, Bonagura et al. Genetically engineered antibodies. Tools for the study of diverse properties of the antibody molecule. Immunol Rev 1992; 130:87-107.
85. Hiatt A, Ma JM-C. Characteristics and applications of antibodies produced in plants. In: Nester EW, Verma DPS, eds Advances in Molecular Genetics of Plant-Microbe Interactions Kluwer Academic Publishers (NL) 1993:549-560.
86. De Wilde C, De Neve M, De Rycke R et al. Intact antigen-binding MAK33 antibody and Fab fragment accumulate in intercellular spaces of *Arabidopsis thaliana*. Plant Sci 1996; 114:233-241.
87. Carpita N, Sabularse D, Montezinos D et al. Determination of the pore size of cell walls of living plant cells. Science 1979; 205:1144-1147.
88. Tepfer M, Taylor IEP. The permeability of plant cell walls as measured by gel filtration chromatography. Science 1981; 213:761-763.
89. Düring K, Porsch P, Fladung M et al. Transgenic potato plants resistant to the phytopathogenic bacterium *Erwinia carotovora*. Plant J 1993; 3:587-598.
90. Krebbers E, Herdies L, De Clercq A et al. Determination of the processing sites of an *Arabidopsis* 2S albumin and characterization of the complete gene family. Plant Physiol 1988; 87:859-866.
91. Hwarig C, Sinskey AJ, Lodish HF. Oxidized redox state of Glutathione in the endoplasmic reticulum. Science 1992; 257:1496-1502.
92. Stieger M, Neuhaus G, Momma T et al. Self assembly of immunoglobulins in the cytoplasm of the alga *Acetabularia mediterranea*. Plant Sci 1991; 73:181-190.
93. Schouten A, Roosien J, van Engelen FA et al. The C-terminal KDEL sequence increases the expression level of a single-chain antibody designed to be targeted to both the cytosol and the secretory pathway in transgenic tobacco. Plant Mol Biol 1996; 30:781-793.
94. Rosso MN, Schouten A, Roosien J et al. Expression and functional characterization of a single chain Fv antibody directed against secretions involved in plant nematode infection process. Biochem Biophys Res Comm 1996; 220:255-263.
95. Bruyns AM, De Jaeger G, De Neve M et al. Bacterial and plant-produced scFv proteins have similar antigen-binding properties. FEBS Lett 1996; 386:5-10.
96. Fecker L F, Kaufmann A, Commandeur U et al. Expression of single-chain anti-

body fragments (scFv) specific for beet necrotic yellow vein virus coat protein or 25K protein in *Escherichia coli* and *Nicotiana benthamiana*. Plant Mol Biol 1996; 32:979-986.

97. Steipe B, Schiller B, Plückthun A et al. Sequence statistics reliably predict stabilizing mutations in a protein domain. J Mol Biol 1994; 240:188-192.

98. Knappik A, Plückthun A. Engineered turns of a recombinant antibody improve its in vitro folding. Prot Engineering 1995; 8:81-89.

99. Knappik A, Krebber C, Plückthun A. The effect of folding catalysts on the in vivo folding process of different antibody fragments expressed in *Escherichia coli*. Bio/technology 1993; 11:77-83.

100. Solar I, Gershoni JM. Linker modification introduces useful molecular instability in a single chain antibody. Prot Engineering 1995; 8:717-723.

101. Pantoliano MW, Bird RE, Johnson S et al. Conformational stability, folding and ligand-binding affinity of single-chain Fv immunoglobulin fragments expressed in *E. coli*. Biochemistry 1991; 30:10117-10125.

102. Alfthan K, Takkinen K, Sizmann D et al. Properties of a single-chain antibody containing different linker peptides. Prot Engineering 1995; 8:725-731.

103. Whitlow M, Bell BA, Feng S-L et al. An improved linker for single-chain Fv with reduced aggregation and enhanced proteolytic stability. Prot Engineering 1993; 6:989-995.

104. Jiang WR, Venugopal K, Gould EA. Intracellular interference of tick-borne flavivirus infection by using a single-chain antibody fragment delived by recombinant Sindbis virus. J Virol 1995; 69:1044-1049.

105. Biocca S, Neuberger MS, Cattaneo A. Expression and targeting of intracellular antibodies in mammalian cells. EMBO J 1990; 9:101-108.

106. Biocca S, Ruberti F, Tafani M et al. Redox state of single chain Fv fragments targeted to the endoplasmic reticulum, cytosol and mitochondria. Bio Technology 1995; 13:1110-1115.

107. Biocca S, Cattaneo A. Intracellular immunization: antibody targeting to subcellular compartments. Trends Cell Biol 1995; 5:248-252.

108. Glockshuber R, Schmidt T, Plückthun A. The disulfide bonds in antibody variable domains: effects on stability, folding *invitro* and functional expression in *Escherichia coli*. Biochemistry 1992; 31:1270-1279.

109. Bent. AF. Plant disease resistance genes: function meets structure. Plant Cell 1996; 8:1757-1771.

110. Baulcombe DC. Mechanisms of pathogen-derived resistance to viruses in transgenic plants. Plant Cell 1996; 8:1833-1844.

111. Tepfer M. Viral genes and transgenic plants. Bio/technology 1993; 11:1125-1132.

112. Hull R, Davies JW. Approaches to nonconventional control of plant virus diseases. Critical Rev Plant Sci 1992; 11:17-33.

113. Carrington JC, Kasschau KD, Mahajan SK et al. Cell-to-cell long-distance transport of viruses in plants. Plant Cell 1996; 8:1669-1681.

114. Saunal H, Witz J, van Regenmortel MHV. Inhibition of *in vitro* cotraslational disassembly of tobacco mosaic virus by monoclonal antibodies to the viral coat protein. J Gen Virol 1993; 74:897-900.

115. Hull R. The movement of viruses in plants. Annu Rev Phytopathol 1989; 27:213-240.

116. Deom CM, Lapidot M, Beachy RN. Plant virus movement proteins. Cell 1992; 69:221-224.

117. Argos P. A sequence motif in many polymerases. Nucleic Acid Res 1988; 16:9909-9916.

118. Gorbalenya AE, Koonin E. Viral proteins containing the purine NTP-binding sequence pattern. Nucleic Acid Res 1989; 17:8413-8434.
119. Williamson VM, Hussey RS. Nematode pathogenesis and resistance in plants. Plant Cell 1996; 8:1735-1745.
120. Baum TJ, Hiatt A, Parrott WA et al. Expression in tobacco of a functional monoclonal antibody specific to stylet secretions of the root-knot nematode. Mol Plant-Microbe Interact 1996; 9:382-387.
121. Knogge W. Fungal infection of plants. Plant Cell 1996; 8:1711-1722.
122. Hiatt A. Antibodies produced in plants. Nature 1990; 344; 469-470.
123. Anand NN, Mandal S, MacKenzie CR et al. Bacterial expression and secretion of various single-chain Fv genes encoding proteins specific for a *Salmonella* serotype B O-antigen. J Biol Chem 1991; 266:21874-21879.
124. Holliger P, Prospero T, Winter G. Diabodies: small bivalent and bispecific antibody fragment. Proc Natl Acad Sci 1993; 90:6444-6448.
125. Pack P, Müller K, Zahan R et al. Tetravalent miniantibodies with high avidity assembling in *Escherichia coli*. J Mol Biol 1995; 246:28-34.
126. Reiter Y, Brinkmann U, Lee B et al. Engineering antibody Fv fragments for cancer detection and therapy: disulfide-stabilized Fv fragments. Nature Biotechnology 1996; 14:1239-1245.

CHAPTER 10

From Phage Libraries to Intracellular Immunization

Nicola Gargano, Luisa Fasulo, Silvia Biocca and Antonino Cattaneo

Based on what has been described and discussed in the previous chapters, the expression and folding ability of antibody domains in ectopic environments (be it *E. coli* or different intracellular compartments) depends in a crucial and still unpredictable way on the primary sequence of the variable regions. The importance of rational, structure-guided or repertoire-based approaches to learn how to improve the stability and the folding of ectopically expressed antibody domains has been discussed. In this chapter we shall describe new ideas and results about how phage display antibody technology can conceivably be exploited to add a new dimension to that of intracellular antibodies.

Phage Derived Antibodies for Intracellular Immunization

All antibodies used so far for intracellular expression have been derived from monoclonal antibodies of predefined specificity. Antibodies are now increasingly derived from repertoires of variable regions displayed on the surface of filamentous phage (see chapter 3). This new source of antibody variable regions presents several advantages over classic monoclonals for the application to intracellular antibody technology.

First, there are a number of "trivial", but important reasons: (i) phage derived antibodies are by definition recombinant. This allows avoidance of problems related to the cloning of hybridoma V regions; (ii) phage derived antibodies are (usually) human, which is important in view of possible therapeutic applications; (iii) phage derived antibodies can be affinity matured by methods dependent on phage display.

None of these reasons represents an intrinsic difference between antibodies derived from monoclonals and antibodies derived from phage. In particular, none of these points is one that represents an insurmountable difficulty that cannot be overcome for monoclonal antibodies by some extra work. For instance, monoclonal antibodies can certainly be cloned from the corresponding hybridoma cell lines even if with some difficulty (see chapter 3). Murine antibodies can be, and are being, "humanized"[1,2] again with some difficulties that can nonetheless be sorted out. Finally, monoclonal derived recombinant antibodies can be turned into a phage format for further engineering, including affinity maturation.

Intracellular Antibodies: Development and Applications, edited by
Antonino Cattaneo and Silvia Biocca. © Springer-Verlag and Landes Bioscience 1997.

All these reasons are, more related to convenience and to practical issues. These are important aspects, however, one could also argue that for some applications, a "classical" monoclonal antibody and its recombinant version may be an unsurpassed reagent.

Another reason that represents an intrinsic difference between mouse-derived and phage-derived antibodies that may be of importance for downstream use as intracellular antibodies. This relates to the fact that only by exploiting the phage display technology it is possible to have at our disposal recombinant polyclonal repertoires of antigen binding antibodies (more generally of antigen binding proteins). This possibility lays the ground for new selection schemes whereby intracellular expression and targeting of antibodies could be exploited:

- to select antibodies on the basis of binding to a given antigen under conditions of intracellular expression (binding selection).
- to select antibodies on the basis of a given function or phenotype provided by the intracellularly expressed antibody (phenotypic selection).

Engineering these repertoires into a format suitable for intracellular expression in eukaryotic cells, and applying appropriate selective pressures, should allow us to isolate new antibody specificities previously unknown, on the basis of the conferred phenotype. In the strict sense this would be a true immunization on the basis of a phenotype conferred by the intracellular antibodies on the cells that express them. Such a strategy could also be employed to select for antibodies suitable for intracellular expression, out of a pool of antibodies with a given antigen specificity, as discussed below.

Intracellular Antibodies and the Yeast Two Hybrid System

We have previously discussed how a good, strong antibody under in vitro conditions does not necessarily perform satisfactorily when expressed intracellularly. In the future it would be desirable to include one selection step to favor the isolation of antibodies binding to a given antigen efficiently under conditions of intracellular expression. We shall describe initial attempts to achieve this by exploiting the yeast two hybrid system.

The two hybrid system[3,4] or interaction trap, provides an experimental system to monitor intracellular (in yeast cells) protein-protein interactions.[5] This system takes advantage of the modular domain structure of eukaryotic transcription factors: transcriptional activation and DNA-binding regions exist as separable domains which need to be physically associated, though not necessarily covalently linked for reconstitution of activity. In the yeast two hybrid system, activation of a reporter construct occurs when the two domains are brought together through the interaction of two polypeptides expressed as activation domain and DNA-binding domain fusion proteins (Fig. 10.1A).

Two hybrid systems all require the expression of these two fusion proteins from specially designed vectors: one encodes the sequence-specific DNA binding domain (yielding the "bait"), such as that from the yeast transcription factor GAL4 or the bacterial repressor protein LexA. The second vector encodes the transcription activation domain usually from either GAL4 or the herpes simplex virus protein VP16 (the "prey"). The yeast strain in which the intracellular interaction assays are carried out includes auxotrophic mutations for the maintenance of the various plasmids involved and appropriately regulated reporter constructs. The

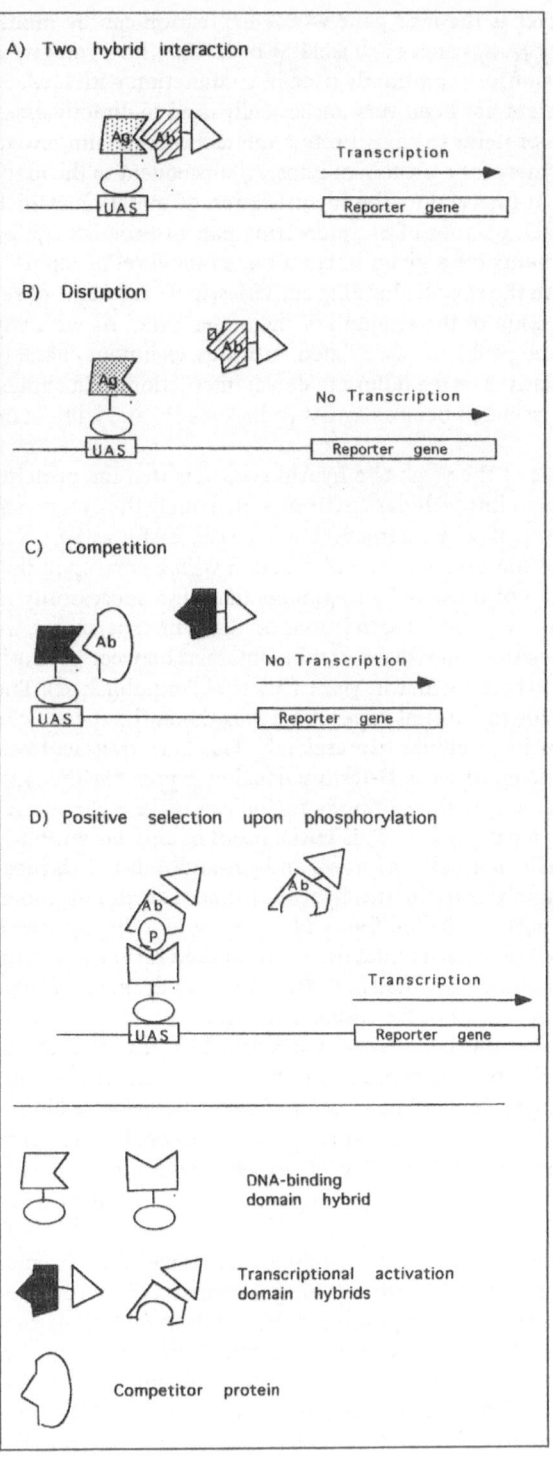

Fig. 10.1. Four hypothetical examples of antibody expression in the two and three hybrid format.

(A) The DNA binding domain fused to the antigen (the bait), specifically binds the upstream activation sequence (UAS) in the reporter construct. The interaction of the activation domain fused to the corresponding ScFv antibody fragment (the prey) with the bait results in the expression of the reporter construct. In (B), a nonrelevant antibody fragment will not give a productive interaction.

(C) In this format, the expression of a ScFv fragment directed against the bait as a third partner, dissociates the bait-prey complex; as in (B), reporter expression is turned off. Growth of yeast colonies where interactions are disrupted can be selected by the use of a counter-selectable marker(toxic reporter gene).

(D) The interaction between the antigen (bait) and antibody (prey) is dependent upon a post-translational modification of the former. This can be carried out by a third partner enzyme.

most commonly used reporter is the lacZ gene whose expression can be monitored by a colorimetric assay. Yeast genes such as HIS3 or LEU2, which can give a growth selection for interaction, are commonly used in conjunction with lacZ.

The yeast two hybrid system has been very successfully used to directly assay interactions between known proteins to study protein interaction domains and to isolate novel interacting partners for a protein of interest. Subsequent to the identification of a protein-protein interaction, the identification of mutations which disrupt the interaction in each partner of an interacting pair of proteins can be useful to probe the requirements for a given interaction. As the level of reporter activation correlates well with the specific binding activities, measurements of reporter activity give an indication of the strength of the interaction.[6] As with any other method, there are some problems associated with this technique, namely the identification of false positives or the failure to detect interactions that should occur (false negatives). This issue has been extensively discussed[5,7] and will not be addressed here.

An intrinsic characteristic of the yeast two hybrid system is that the protein-protein interactions occur in an intracellular environment, namely the cytoplasm and the nucleus. For this reason, if ScFv fragments could be successfully expressed in a yeast two hybrid format, monitoring their interaction with a corresponding antigen should allow isolation of those ScFv fragments that bind successfully to the antigen under intracellular expression conditions. In a recent experiment, we have demonstrated that a positive interaction can be obtained between an antigen-antibody pair in a two hybrid format in yeast (AC et al., unpublished). The particular antibody chosen for this initial experiment was the anti p21-ras ScFv fragment previously used for intracellular expression.[8,9] This ScFv fragment was fused to the GAL4 activation domain as an N-terminal fusion to preserve the antigen binding site free for interaction; this antibody fusion can make a productive and specific interaction with a p21-ras bait. This initial proof of antigen-antibody interaction in the two hybrid format, using a known antigen-antibody pair, is presently being extended to the analysis in the two hybrid format of small polyclonal repertoires selected from phage display antibody libraries by panning against a given antigen. In this format, the experimental design is tailored for the selection of a polyclonal population of affinity purified SvFv fragments on the basis of their binding to a given antigen, regardless of functional consequences. This would be the antibody equivalent of what has been recently described for the selection of binding partners from peptide libraries expressed in the two hybrid format. This has been done either by fusing to the GAL4 activation domain, the peptide library directly[10] or the peptide library displayed as part of *E. coli* thioredoxin.[11] The major application of such a scheme would be that of preselecting antigen-specific ScFvs before their subsequent validation and exploitation in functional intracellular antibody experiments.

Recent developments in the two hybrid field allow one to envisage other schemes that would broaden the range of applications. Using a reverse two hybrid approach,[12-14] a positive selection can be imposed for dissociating mutations in a given interacting pair (Fig. 10.1C). This involves the yeast URA3 gene as a reporter. Not only is the URA3 gene product essential for uracil biosynthesis, it also catalyzes the conversion of an artificial substrate (5-fluoroorotic acid, 5-FOA) into a toxic compound. Thus, an interaction can be selected against, or conversely, the dissociation of an interaction by mutation or by a competing protein (Fig. 10.1C) can be selected favorably. Exploiting the reverse two hybrid approach, the counterselection could be used to identify ScFvs that compete with and dissociate a specific protein-protein interaction. In this case, antibodies would not be expressed as one of the two hybrid fusion partners, but as third competing partners expressed as nuclear targeted proteins.

Protein-protein interactions are often dependent on the post-translational modification of one component of the complex. In a recently described application of the two hybrid system, the so-called yeast tribrid,[15] a nonproductive interaction between a "two hybrid" protein pair which depends on one of the partners being phosphorylated, is turned on by the expression of the relevant kinase as a third partner. Antibodies specific for post-translationally modified proteins could conceivably be used along these lines (Fig. 10.1D) to facilitate the expression cloning of unknown enzymes carrying out a given post-translational modification of a target protein.

Intracellular Libraries in Mammalian Cells

Intracellular Expression of a Phage Derived Antibody Fragment in Mammalian Cells: A Proof of Principle

Before describing how selection strategies based on the phenotype conferred by intracellular antibody fragments could be implemented, it is necessary to ask whether antibody fragments suitable for intracellular selection can be isolated from phage display libraries and whether these can be effective as intracellularly expressed proteins. In particular, it remains to be verified whether the affinity of a phage derived antibody fragment is high enough to achieve an effective intracellular neutralization of the corresponding antigen.

The first "proof of principle" demonstrating the successful use of a phage derived antibody domain for intracellular expression in mammalian cells has been recently provided.[16-18]

An antibody fragment directed against the reverse transcriptase (RT) of HIV was isolated[16] from a synthetic combinatorial library of human Fab antibody fragments displayed on the surface of filamentous phage,[19] using recombinant HIV-1 RT as a solid phase selector. Successive cycles of binding, elution and phage amplification were performed. At each cycle, the enrichment in phages able to bind RT and to inhibit the RNA-dependent DNA polymerase activity of RT was assessed (Fig. 10.2). Individual phage clones were further characterized and one of these antibody fragment, D7 was selected for further applications. This antibody fragment neutralizes very efficiently the RNA-dependent DNA polymerase activity of RT at nanomolar concentrations. The inhibitory activity of the anti-RT antibody fragment, D7, is competitive with respect to the template primer. The antibody fragment has a broad spectrum of reactivity with respect to other polymerases of the RT family, since it also neutralizes the activity of RT from avian and murine retroviruses in particular Moloney Mouse Leukemia Virus (MMLV). Moreover, the D7αRT fragment also inhibits the RNA-dependent DNA-polymerase activity of HIV-1 drug resistant RT mutants (Leu74→Val, dideoxyinosine resistant, and RTMC, AZT resistant), as well as that of RT from HIV-2 and SIV. This antibody fragment, therefore, appears to recognize a structural motif important for the enzymatic activity and conserved among RT enzymes only distantly related in terms of primary sequence.[16] In this respect, the broad reactivity properties of the antibody fragment D7 demonstrate that phage display libraries may provide binding specificities not readily obtainable with classical monoclonal antibodies.

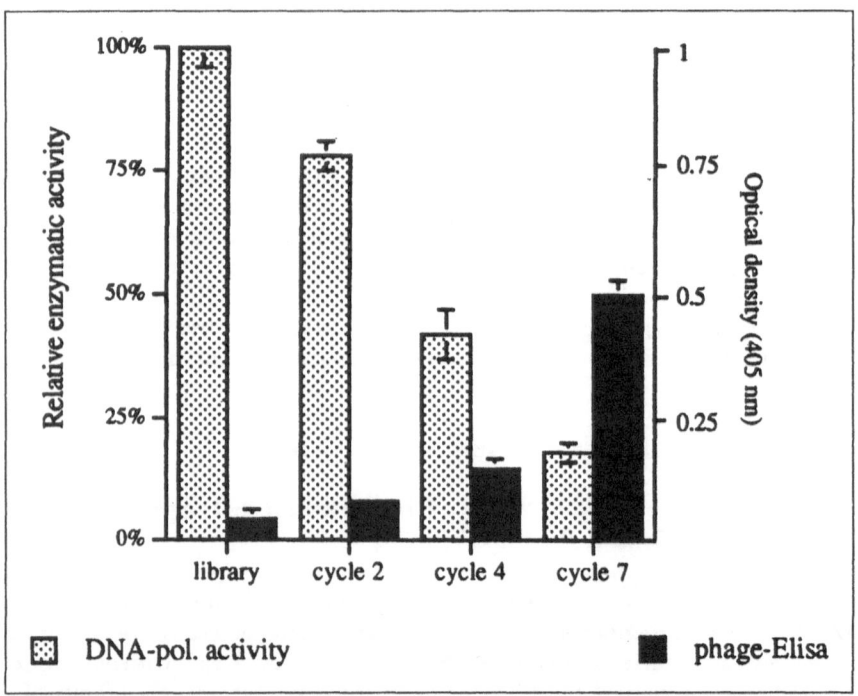

Fig. 10.2. Selection of HIV-1 RT neutralizing antibody fragments from a phage-display library. Polyclonal phage-antibodies after each cycle of selection were assayed by ELISA for binding to p51/66 RT in solid phase (filled bars). Bound phages to solid phase were detected with an anti-fd sheep antiserum followed by anti-goat peroxidase conjugate antibodies. The same polyclonal phage-antibodies, at subsequent cycles of selection, were assayed for their ability to inhibit the RNA-dependent DNA-polymerase activity of recombinant HIV-1 RT. p51/66 RT was incubated in the presence of phage-antibodies derived from the unselected library and with the "polyclonal" phages after two, four and seven rounds of RT selection. These mixtures were then assayed for RT RNA-dependent DNA-polymerase activity under standard conditions.

On the basis of a detailed structural comparison between MMLV and HIV-1 RT,[20] it has been proposed that the structure of MMLV RT may provide a model for understanding mutations in HIV-1 RT associated with nucleoside analog resistance found in clinical isolates. That drug design could be facilitated by targeting conserved regions of the protein and testing for a drug's ability to inhibit MMLV RT as well as HIV-1 RT. Following this line of reasoning, the ability of the D7aRT antibody fragment to inhibit Moloney MLV retrotranscription, upon intracellular expression, was investigated.

The D7αRT antibody fragment was expressed as a cytosolic protein Balb/c-3T3 cells, and transfected cells were challenged with two Moloney-based, replication defective, recombinant viruses, (MMLV-βgal and MMLV-HSVtk), which drive the expression of the reporter genes for *E. coli* β-galactosidase and for herpes simplex virus thymidine kinase respectively. The latter gene confers sensitivity to the toxic

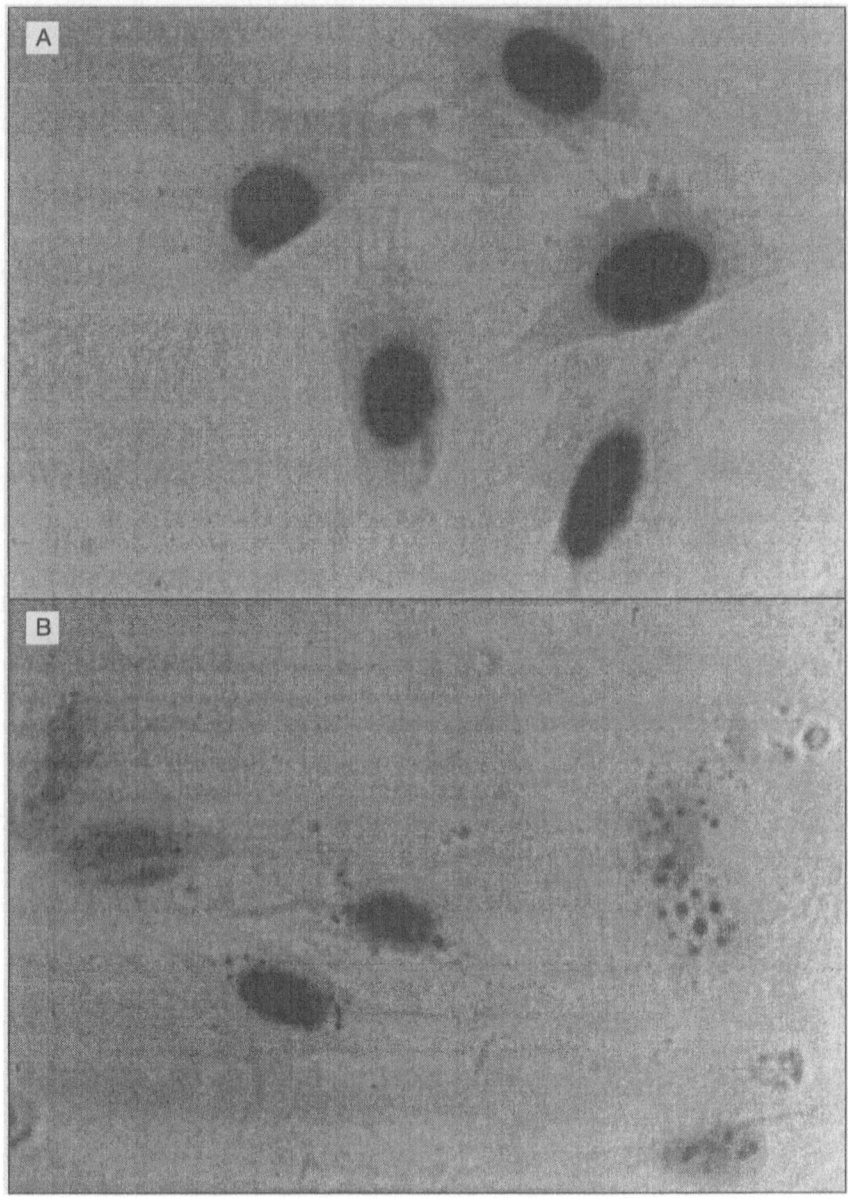

Fig. 10.3. β-galactosidase staining in cells expressing D7αRT. Balb/c-3T3 cells expressing cytoplasmic anti-NGF (panel A) and D7αRT (panel B) antibody fragments were infected with the replication defective MMLV-βgal virus and stained 3 days after the infection for the expression of the retrovirally encoded reporter gene β-galactosidase. Magnification 40X.

action of Gancyclovir (GCV). Figure 10.3 shows that the β-galactosidase staining of MMLV-βgal-infected cells is greatly reduced in cells expressing the D7αRT antibody fragment, as a result of a reduced efficiency of viral retrotranscription. The same cell transfectants were infected with the MMLV-HSVtk retrovirus and cultured with GCV. The expression of the HSV-tk gene converts the drug Gancyclovir (GCV) into a toxic form, thus leading to GCV-dependent cell death. As shown in Figure 10.4A, the intracellular expression of the D7αRT antibody fragment leads to a selective survival of the cells which are not affected by the toxic effects resulting from the MMLV-HSVtk retrovirus infection. The initial phase of cell death reflects the nonclonal nature of the transfected population and the following growth phase represents the expansion of the more resistant cells in the initial population. This experiment demonstrates a cell selection on the basis of the function of the intracellularly expressed antibody fragment: the cell resistance to the toxic effect of the virally expressed HSV-tk correlates with an increased expression of the antibody fragment in the cell population (Fig. 10.4B). Moreover, these results provide a demonstration that the intracellular expression of an antibody fragment derived from a phage library effectively inhibits retroviral retrotranscription.

Rescueing Cells on the Basis of the Phenotype Conferred by an Intracellular Antibody

Model selections were performed using resistance to the toxic action of the MMLV-HSVtk retrovirus as the selected phenotype. At first, these were performed, by diluting cells expressing the D7aRT antibody fragment with an excess of negative cells. The cell population was infected with the MMLV-HSVtk retrovirus and cultured with GCV. The cells expressing the D7aRT antibody fragment can be efficiently rescued and expanded within a background heterogeneous cell population using intracellular expression of the antibody fragment as the sole selectable marker. As a result, the cell population is greatly enriched in the antibody fragment (Fig. 10.7). In another experiment, D7aRT DNA was "diluted" up to 1/1000 with DNA encoding for a polyclonal population of nonrelevant antibody fragments. Cells transfected with this DNA mixture were challenged with MMLV-HSVtk retrovirus and GCV. It was also found that the number of surviving clones correlates with the input D7 DNA in the transfection mixture and cells expressing the D7aRT DNA are enriched, as the selection proceeds, at the expense of cells expressing other, less efficient, antibody fragments.

In these model selections, the neutralizing properties of the D7aRT antibody fragment were known beforehand, but in principle this selection procedure exploiting intracellular expression and a selectable phenotype could be used to rescue cells expressing a previously unknown neutralizing antibody fragment against a polyclonal background of non-neutralizing ones. From these cells, the corresponding intracellularly expressed antibody fragment could be isolated and characterized. In a way, this would represent a sort of functional epitope mapping in vivo. Thus, these results provide a "proof of principle" for an experimental strategy, whereby a library of antibody fragments would be intracellularly expressed and exposed to the selective pressure of a cytotoxic virus. It is clear that the success of this scheme will depend strongly on the tightness and the rapidity of the selection scheme.

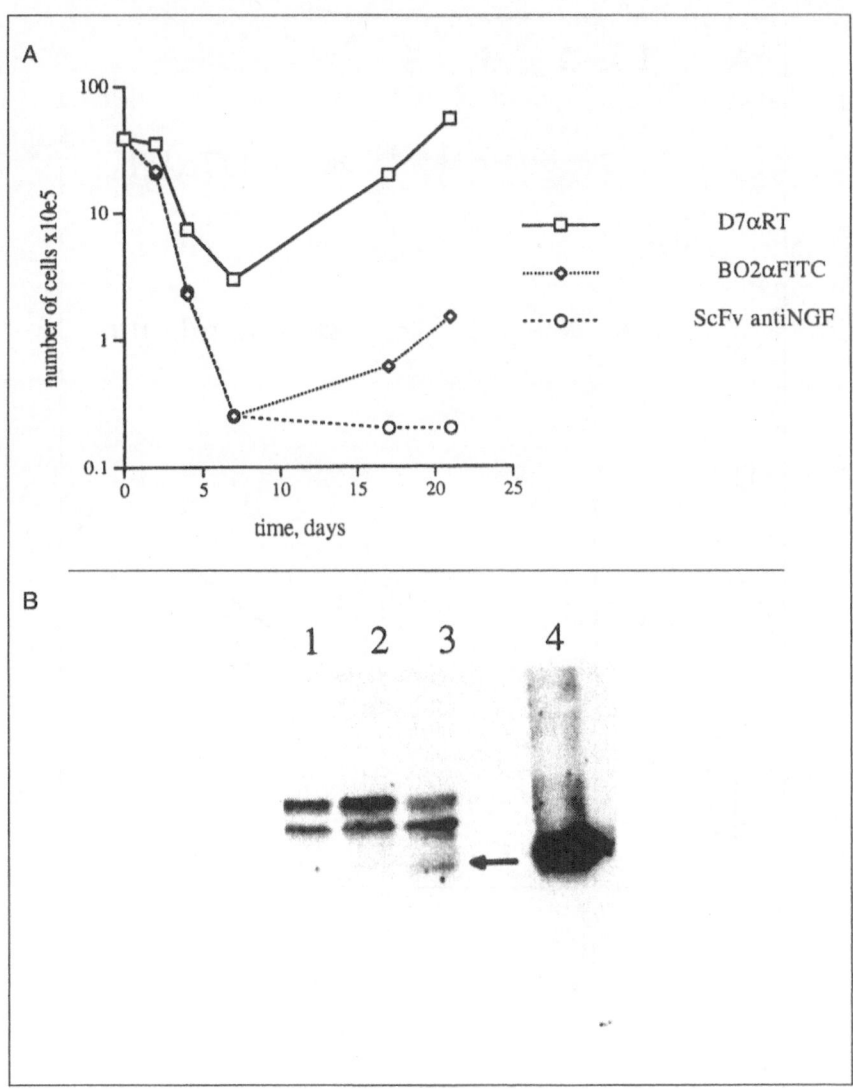

Fig. 10.4. (A) Resistance of D7αRT expressing fibroblasts to the toxic action of retrovirally encoded herpes simplex virus thymidine kinase (HSV-tk). Balb/c-3T3 cells transfected with D7αRT and with control antibody fragments (BO2αFITC and ScFv-anti-NGF) DNA, were infected with the replication defective MMLV-HSVtk retrovirus (PAGO). 24 hours later (day 0) the medium was replaced and 10 mg/ml gancyclovir (GCV) was added. The number of surviving cells was determined (quadruplicates) at different time points during GCV selection. (B) Enrichment of D7αRT expressing cells after PAGO infection and GCV selection. Uncloned D7αRT transfected Balb/c-3T3 cells were exposed to the HSV-tk (PAGO) killing selection shown in A). At different times of GCV selection, 2.5×10^5 cells were collected (day 0 - lane 1; day 4 - lane 2; day 15 - lane 3), and assayed for the expression of D7αRT antibody fragments by Western analysis. Lane 4: equivalent amounts of D7αRT baculovirus infected insect cells.

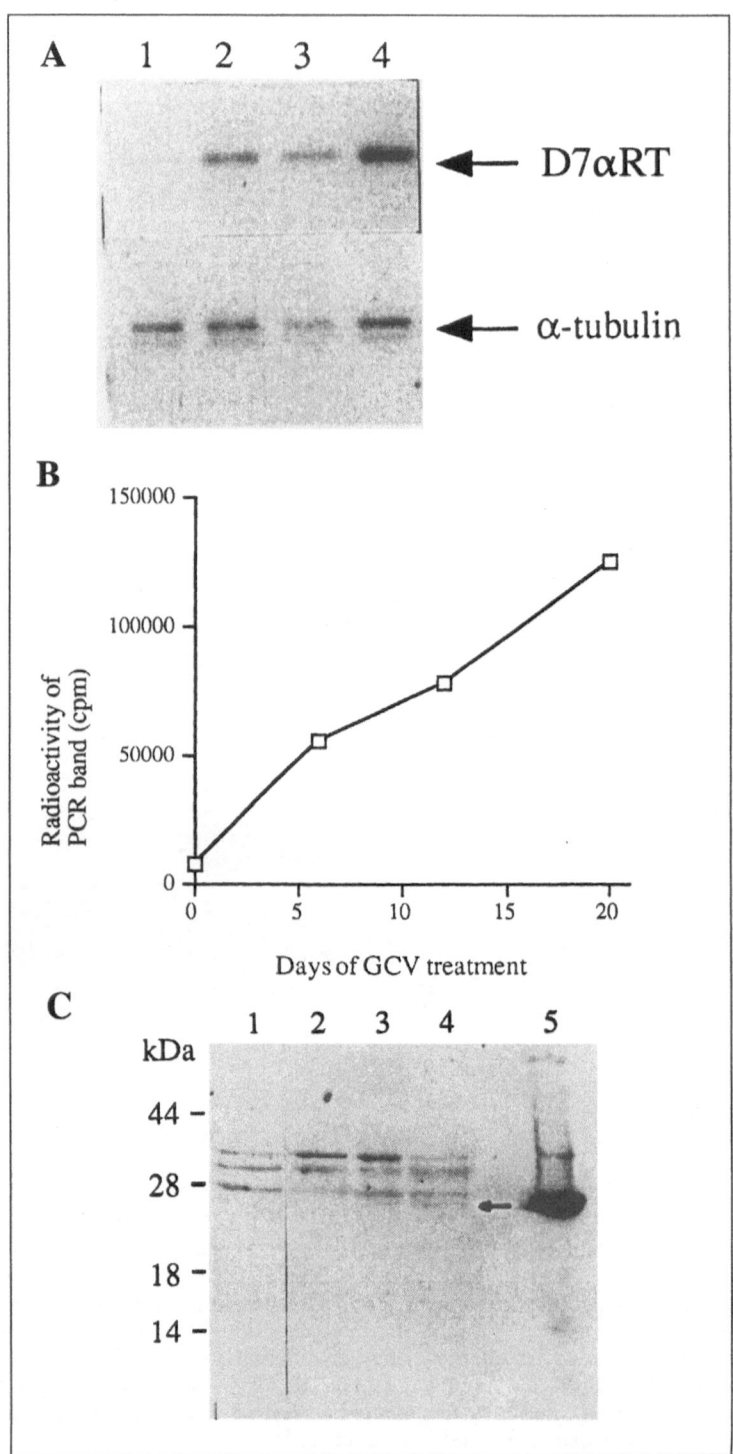

More generally, provided a suitable selection scheme is designed, this should allow the rescue of antibody specificities providing a selection advantage to the cells expressing them intracellularly. By way of example, we illustrate this strategy for the case of resistance to a cytotoxic virus (Fig. 10.6). A phage display library of antibody fragments is challenged with solid-phase coupled antigen (Ag), reverse transcriptase (RT) in this example, and an affinity purified population of phages enriched in anti-Ag specificities is obtained after a few cycles. Only a few of these specificities, however, will be "neutralizing" ones. Bypassing the in vitro characterization of individual antibodies in this polyclonal population, this small enriched repertoire is formatted for intracellular expression in mammalian cells using vectors now available for this purpose[21] and is transfected as a pooled DNA. Individual cells will express a small subset of antibody specificities or even single specificities. Intracellular expression can be used to favorably select and rescue the cells expressing the neutralizing antibodies. These cells will outgrow the culture and will allow the isolation of the particular antibody expressed.

The assay described lends itself in a natural way to select out of a polyclonal pool of antibody specificities, intracellular antibody fragments more efficient than others in providing the phenotype necessary for the cell to survive an imposed selective pressure. This is not limited to the one described; other schemes of phenotypic selection could be devised in which the intracellular expression of antibodies confers growth advantage to cells expressing neutralizing antibody specificities (e.g. resistance to the action of apoptotic proteins).

The nature of the polyclonal repertoire need not be restricted to the example given, namely a polyclonal mixture of antibodies derived from a partial selection of a phage library on an antigen column. For instance, the polyclonal repertoire may be represented by a pool of mutated versions of one given lead antibody fragment, possibly made by expression in an *E. coli* mutator strain or by chain shuffling (see chapter 3). This would allow selection for higher affinity or for improved performance to be performed in vivo, although the actual feasibility of this remains to be verified.

Fig. 10.5 (opposite). Rescue of D7αRT expressing cells under gancyclovir selective pressure. D7αRT expressing cells were mixed at a ratio of 10^{-2} with untransfected Balb/c-3T3 cells and infected with PAGO viruses. At different times after PAGO infection and GCV exposure, cells were collected and assayed for the levels of D7αRT protein and DNA.

(A) Southern analysis of PCR products. Total cellular DNA prepared from 10^4 cells was amplified by PCR with primer pairs complementary to the D7αRT gene and to α-tubulin cellular gene. Cell samples derived before gancyclovir treatment (lane 1) after 6 (lane 2), 12 (lane 3) and 15 days (lane 4) of GCV exposure were analyzed (the upper arrow refers to the D7aRT band, the lower one to α-tubulin).

(B) Quantitative analysis of D7αRT DNA. The levels of anti-RT antibody fragment DNA were normalized to α-tubulin and plotted as amount of radioactivity in the hybridized bands (Phosphorimager).

(C) Western analysis of D7αRT antibody fragments during GCV selection. 2.5×10^5 cells were assayed before gancyclovir exposure (lane 1), after 6 (lane 2), 12 (lane 3), or 15 days of GCV treatment (lane 4), and visualized with mAb 9E10 by ECL. Equivalent cell extracts of D7αRT baculovirus infected insect cells (lane 5) were analyzed as control.

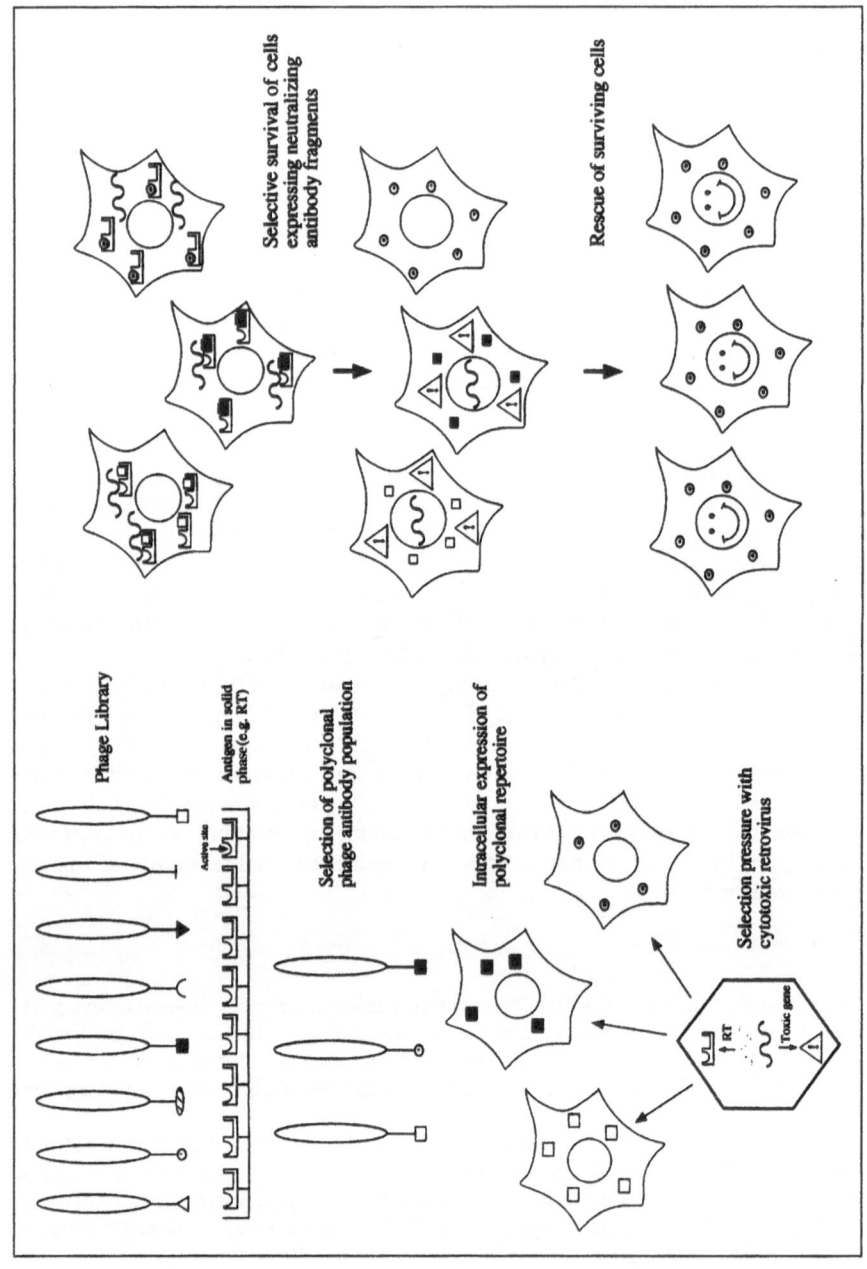

Fig. 10.6. Schematic view of the in vivo selection of neutralizing anti-RT antibody fragments.

It is clear that as we are dealing with mammalian cells, the size of the repertoire cannot be too large—hence the need of a preselection step such as affinity purification on an antigen column. However, the polyclonal repertoire can also be restricted in alternative ways. Thus, a number of alternative procedures to enrich in specific binders of interest from a large initial library of antibodies could be used, (e.g., for instance, phage display subtraction methods for the competitive enrichment of antibodies directed against proteins constituting a difference between two populations of cells).[22] In this method, the target proteins from one cell population are coupled to a solid phase and the competitive protein extract from the second cell population is in solution and competes with binding to the solid phase. This results in a polyclonal library enriched for specificities present in one of the two competing protein extracts.

With antibodies derived from large phage display libraries, the intracellular targeting of "polyclonal" repertoires of antibody domains is now feasible and provided a suitable selection procedure is applied, this should allow us to select specific antibodies against known or unknown antigens (or epitopes), or with improved properties on the basis of a selectable phenotype: a true form of intracellular immunization.

References

1. Jones PT, Dear PH, Foote J et al. Replacing the complementarity determining regions in a human antibody with those from a mouse. Nature 1986; 321:522-525.
2. Riechmann L, Clark M, Waldmann H et al. Reshaping human antibodies for therapy. Nature 1988; 332:323-327.
3. Fields S, Song O. A novel genetic system to detect protein-protein interactions. Nature 1989; 340:245-247.
4. Chien CT, Bartel PL, Sternglanz R et al. The two hybrid system: a method to identify and clone genes for proteins that interact with a protein of interest. Proc Nat Acad Sci USA 1991; 88:9578-9582.
5. Fields S, Sternglantz R. The two hybrid system: an assay for protein- protein interactions. Trends Genet 1994; 10:286-292.
6. Estojak J, Brent R, Golemis EA. Correlation of two hybrid affinity data with in vitro measurements. Mol Cell Biol 1995; 15:5820-5829.
7. Finley RL, Brent R. Interaction trap cloning with yeast. In: Hames BD, Glover DM, eds. DNA Cloning 2, Expression Systems: A Practical Approach. Oxford: Oxford University Press, 1995:169-203.
8. Biocca S, Pierandrei-Amaldi P, Cattaneo A. Intracellular expression of anti-p21ras single chain Fv fragments inhibits meiotic maturation of *Xenopus* oocytes. Biochem Biophis Res Commun 1993; 197:422-427.
9. Biocca S, Pierandrei-Amaldi P, Campioni N et al. Intracellular immunization with cytosolic recombinant antibodies. Bio/Technology 1994; 12:396-399.
10. Yang M, Wu Z, Fields S. Protein-peptide interactions analyzed with the yeast two hybrid system. Nucleic Acid Res 1995; 23:1152-1156.
11. Colas P, Cohen B, Jessen T. Genetic selection of peptide aptamers that recognize and inhibit cyclin-dependent kinase 2. Nature 1996; 380:548-550.
12. Vidal M, Brachmann RK, Fattaey A et al. Reverse two hybrid and one hybrid systems to detect dissociation of protein-protein and DNA- protein interactions. Proc Nat Acad Sci USA 1996; 93:10315-10320.
13. Vidal M, Braun P, Chen E et al. Genetic characterization of a mammalian protein-protein interaction domain by using a reverse two- hybrid system. Proc Nat Acad Sci USA 1996; 93:10321-10326.

14. White MA. The yeast two hybrid system: forward and reverse. Proc Nat Acad Sci USA 1996; 93:10001-10003.
15. Osborne MA, Dalton S, Kochan JP. The yeast tribrid system: genetic detection of trans-phosphorylated ITAM-SH2-interactions. Bio/Technology 1995; 13:1474-1434.
16. Gargano N, Biocca S, Bradbury A et al. Human recombinant antibody fragments neutralizing human immunodeficiency virus type 1 reverse transcriptase provide an experimental basis for the structural classification of the DNA polymerase family. J Virol 1996; 70:7706-7712.
17. Gargano N, Cattaneo A. From phage libraries to intracellular antibodies. Proceedings of the IBC Seventh Annual International Conference on Antibody Engineering 1996; in press.
18. Gargano N, Cattaneo A. Inhibition of Murine Moloney Leukemia Virus retrotranscription by the intracellular expression of a phage-derived anti-reverse transcriptase antibody fragment. J Gen Virol 1997; in press.
19. Griffiths AD, Williams SC, Hartley O et al. Isolation of high affinity human antibodies directly from large synthetic repertoires. EMBO J 1994; 13:3245-3260.
20. Georgiadis MM, Jessen SM, Ogata CM et al. Mechanistic implication from the structure of a catalytic fragment of Moloney leukemia virus reverse transcriptase. Structure 1995; 3:879-892.
21. Persic L, Righi M, Roberts A et al. Targeting vectors for intracellular immunization. Gene 1997; 187:1-8.
22. Stausbol-Gron B, Wind T, Kjaer S et al. A model phage display subtraction method with potential for analysis of differential gene expression. FEBS Lett 1996; 391:71-75.

CHAPTER 11

Perspectives and Conclusions

Antonino Cattaneo and Silvia Biocca

The technology of intracellular antibodies and more generally of ectopic antibody expression, has a great potential in many different fields. The work performed so far has indicated quite clearly which are the directions in which improvements are needed and which are the points that need to be taken care of.

The affinity of the antibodies is of course an important requisite for their successful use. While the phage technology now allows to isolate recombinant antibodies in a form suitable for their ectopic expression, it also provides the technology for an improvement of their affinity in vitro. Besides this form of affinity maturation in vitro, there are other means to increase the "effective affinity" of the antibodies. The choice of the antibody form allows to exploit multivalency of binding (avidity effect) to strengthen the binding interactions. Future work will tell us if and how multivalent minibodies or CRABs will lead to an improved efficacy for intracellular expression over the monovalent ScFvs.

The stability of the intracellularly expressed antibodies is an issue of great importance as it affects their half-life in the cell. This problem is particularly relevant for antibody fragments expressed in the reducing environment of the cell cytoplasm. We envisage that two lines of research will lead to a solution of this problem. The first one is the rational study of the structural requirements for the folding of antibody domains expressed in *E. coli* or in other artificial environments (such as the test tube). The second one will exploit selection schemes whereby antibodies are also selected on the basis of their stability when ectopically expressed in vivo. Accordingly, intracellular antibodies may be selected on the basis of their binding to a given antigen or on the basis of a phenotype conferred to cells in which they are expressed. It is anticipated that these studies will lead to variable region frameworks more suitable for intracellular expression.

A precise targeting of the antibodies close to the intracellular location where their corresponding antigen is active, or diverting it from reaching its final destination is the crucial aspect of this technology. Targeting of functionally active antibodies to quite a few intracellular compartments has been achieved, but further refinements of the targeting strategies are conceivable, exploiting the growing knowledge of intracellular traffic signals. Some intracellular traffic pathways can be followed in two directions. For instance, the exocytic/endocytic pathways, or the retrotranslocation of proteins from the ER back into the cytoplasm. The use of these reverse pathways has not been exploited for intracellular antibodies yet.

Intracellular Antibodies: Development and Applications, edited by
Antonino Cattaneo and Silvia Biocca. © Springer-Verlag and Landes Bioscience 1997.

In its simpler scheme, the intracellular antibodies act exclusively as binders. Neutralization of the activity of the bound antigen may follow. Antibodies normally carry in their constant domains, effector functions that are relevant for their function in the immune system, but ectopically expressed antibodies do not require such effector functions. Their usual format for this application is that of small binding units, for instance ScFv fragments. New effector functions can be added to the binding moiety. Anchoring sequences preventing the recognized protein from reaching the cellular location where it is active, are the simpler example of an effector function (intracellular anchors). In this mode of action, antibodies that are not neutralizing can be effectively used to inhibit the function of the target protein. In a more general way, effector functions would allow to change the mode of action of intracellular antibodies, for instance, turning them into vectors for intracellular degradation or into intercellularly acting immunotoxins. One attractive possibility for the future would be to use a natural or engineered fluorescent protein (such as the jelly fish green fluorescent protein or mutants derived) as an effector function. If conditions could be found, whereby binding to the antigen leads to a change in the fluorescence properties, this could be used for imaging protein-protein interaction in a living cell.

The intracellular antibody strategy increases the spectrum of the antigens which are accessible to the technology of therapeutic antibodies presently limited to the accessibility of therapeutically relevant antigens in the extracellular environment. One exciting application of the technology of ectopic antibody expression for intracellular and intercellular immunization is for somatic gene therapy in human disease (for instance, HIV infection). It is clear that specific problems to be solved, in that perspective, relate to aspects in common with other candidate genes for gene therapy. In particular, the delivery of the gene to the target cells and the immunogenicity of the therapeutic gene. The antigen-presentation of proteolytically processed intracellular antibodies is still an open issue. It is anticipated that the clinical trials presently under way for intracellular antibodies will start shedding some light on these aspects.

The applications of ectopic antibody expression in plants are particularly promising in terms of engineering resistance to pathogens and for passive immunization of humans. The latter is a form of intercellular immunization, in the strict sense, across species. This concept is likely to be extended in the future to immunize plants against parasitic infections.

Antibodies are increasingly being derived from large phage display libraries. One problem in common with all repertoires, be it chemical, peptides, RNA or antibodies, is that of accessibility and of selection procedures. An area likely to be expanded in the future is that of the intracellular expression of antibody repertoires. The concept of in vivo selection of antibodies on the basis of a selectable phenotype, may provide new ways to access the repertoire diversity, and may be developed in *E. coli*, yeast or even mammalian cells with different formats and different purposes. One application relates to the potential of performing what we call "in vivo epitope mapping". This is notable since (i) for many protein antigens, an in vitro neutralization assay is not available and an active epitope is not known a priori, (ii) the efficacy of an intracellularly targeted antibody fragment does not always correlate with properties of the parental antibody, such as affinity or specificity as studied in vitro and (iii) the folding and stability properties of ectopically expressed antibodies is an important and still unpredictable variable affecting the performance of intracellular antibodies, other things being equal.

In general, cells are well-suited for high-throughput screening procedures. Intracellular expression of antibody repertoires in schemes where the desired antibodies provide a growth advantage, could provide a new class of cell-based assays for the isolation of lead antibodies and targets for drug design. It is obvious that such schemes can only be implemented by exploiting the phage technology: only with phage technology is it possible to have the recombinant version of "classical" polyclonal antisera since ScFv fragments are coselected together with their corresponding genes.

In conclusion, the strategy of phenotypic selection of intracellular repertoires represents a meeting point between the technologies of phage antibody display and of intracellular antibodies, a true form of intracellular immunization.

INDEX